FACTS AT YOUR FINGERTIPS

INVENTION AND TECHNOLOGY
MILITARY AND SECURITY

Published by Brown Bear Books Ltd
First Floor
9-17 St. Albans Place
London N1 0NX
UK

© 2012 Brown Bear Books Ltd

Library of Congress Cataloging-in-Publication Data

Military and security / edited by Tom Jackson.
    p. cm. – (Facts at your fingertips. Invention and technology)
  Includes index.
 Summary: "Describes the evolution of warfare and weapons from pre-Civil War to today's modern techniques on land and sea. Also describes the changes in security devices, such as locks and keys. A timeline traces the history of warfare"–Provided by publisher.
  ISBN 978-1-936333-37-0 (library binding)
 1. Military art and science. 2. Locks and keys. I. Jackson, Tom, 1972-

 U105.M55 2012
 355.02–dc23

2012004788

All rights reserved. This book is protected by copyright.
No part of it may be reproduced, stored in a retrieval system, or transmitted in any form or by any means, without the prior permission in writing of the publisher, nor be otherwise circulated in any form of binding or cover other than that in which it is published and without a similar condition including this condition being imposed on the subsequent publisher.

Editorial Director: Lindsey Lowe
Editor: Tom Jackson
Creative Director: Jeni Child
Designer: Lynne Lennon
Children's Publisher: Anne O'Daly
Production Director: Alastair Gourlay

Printed in the United States of America

CPSIA compliance information: Batch #AG/5509

Picture Credits

Front Cover: US Department of Defense
Back Cover: US Department of Defense

Getty Images: De Agostini Picutre Library 45b; SSPL 54t **Public Domain**: 41b, 43; James F. Gibson 18t; Library of Congress 53b **Robert Hunt Library**: 21, 29t **Shutterstock**: Algol 8t; Denis Barbulat 8b, 11b; Ryan Rodrick Beiler 3; BMCL 32t; Bryan Busovicki 26b; Diego Cervo 27br; Deon 27bl; Ensuper 49t; ES3D Studios 49b, 61b; Jean-Michel Girard 20b; GLYPHstock 27t; GoodMood Photo 52b; Mark Graves 53t; Michael Hare 40; Van Hart 47; Jumpingsack 25t, 60br; Kaspri 59t; Levent Konuk 58t; Yu Lan 12r; Litvin Leonid 6r; Lorna 59b; Robyn Mackenzie 55b; William Attard McCarthy 39tr; Regien Paassen 4-5; Russell Shively 23b; stocker 42-43; John Wollwerth 37bl; yuri 36tl **Thinkstock**: Comstock 56b; Digital Vision 39b; Hemera 4bl, 2l, 16, 24, 25b, 26t, 29b, 33, 48t Istockphoto 5, 6l, 7t, 10tl, 19, 22t, 28, 32b, 34b, 42bl, 45t, 52t, 56t, 60t, 61t Lifesize 34t; Photodisc 30; Photos.com 13, 14-15, 18b, 22b, 23, 44, 46, 48b **Topfoto**: HIP 15t; The Granger Collection 41t **US Department of Defense**: 1, 35, 51t

Brown Bear Books has made every attempt to contact the copyright holder. If you have any information please email smortimer@windmillbooks.co.uk

All artwork copyright Brown Bear Books Ltd

CONTENTS

Warfare Before the Civil War	4
Civil War to the Nuclear Age	18
Modern Warfare	30
Sea Warfare	40
Locks and Keys	52
Timeline	*60*
Glossary and Further Resources	*62*
Index	*64*

WARFARE BEFORE THE CIVIL WAR

Wars have been fought over the control of land for as long as people have walked the Earth. In every battle, the ingenuity and effectiveness of a force's weapons can make the difference between winning and losing.

The first battles were probably fought with hunting implements such as as spears and stone daggers. The earliest weapon to be designed specifically for the purpose of warfare is thought to be the club, or mace. The earliest maces, made at the beginning of the Bronze Age (10,000 B.C.), consisted of a stone attached to a wooden handle, and were designed to crush the skull of an enemy. In the third millennium B.C. advances in metalworking allowed some mace heads to be made of copper, and the greater density of the metal transformed the mace into a more effective weapon.

As offensive weapons grew more advanced, so did defensive armor. By 2500 B.C. smiths from Sumeria (modern-day Iraq) were able to craft fairly sophisticated helmets out of bronze, and their effectiveness forced changes in the design of the mace. In response, the head of the mace became more oval in shape, so that the force of a blow was concentrated on a specific point, causing greater damage to the helmet. Gradually the mace evolved into the battleax. Unlike the mace, the ax was not merely reliant

THE SAMURAI SWORD

The first iron swords were made about 3,200 years ago. Swords have taken many forms since then, but many believe that the art of the swordsmith reached its high point in the *katana*, the weapon of the Japanese samurai warriors. Before it was placed in a furnace, the katana's blade was coated in clay apart from the cutting edge. This unique process allowed the sword to combine extreme flexibility with a razor-sharp blade.

▶ A samurai sword had one cutting edge that had to be cleaned and oiled regularly to prevent it from rusting and becoming blunt.

MILITARY AND SECURITY

▲ *Primitive cutting blades were made from stone, such as this Native American arrowhead.*

on its weight to inflict damage but used a cutting edge as well.

The ax remained the most important cutting-edge hand weapon until around 1200 B.C., when swords first began to be cast out of iron. While swords had been made before this date, they were cast from bronze, which could not be made into an effective strong, long, and

▼ *Roman soldiers used the testudo—or tortoise—formation during sieges. The platoon is protected from attack from above and in front by the leather-covered shields.*

ARMOR

As early as 1500 B.C. the Egyptians were making body armor by covering leather clothing with small pieces of bronze. The Assyrians were the first to use iron to make such armor. Greek artifacts from the 3rd century B.C. provide the earliest evidence of mail, flexible body armor made of small overlapping or linked metal pieces. There is some evidence that mail was used even earlier in Celtic Britain. By the 1100s A.D. suits of mail armor covered the entire body. Improvements in the metalworking skills of European armorers in the 13th century resulted in the production of cheaper and stronger iron. As a result, chain mail was gradually replaced with plate armor, which was made from larger pieces of metal. At first plate and mail were used together. However, the gaps between the plates created vulnerable spots and led to the creation of complete suits of armor in Germany and Italy in the 14th century.

◀ *A history enthusiast tests out fighting in plated armor.*

WARFARE BEFORE THE CIVIL WAR

sharp blade. It was only after the development of iron smelting that the sword came into its own. While its design changed repeatedly over the next 3,000 years, it was used successfully in battle right up until the late 19th century.

Slings and arrows

The sword, ax, and mace are all shock weapons, which are used in hand-to-hand combat. Equally important were missile weapons, which were used to attack the enemy from afar. The simplest of these was the sling, which consisted of two thongs attached to a pouch. A small stone or lead shot was put in the pouch. The thrower whirled the sling around his head before

> **FACTS AND FIGURES**
>
> - The longbow was 6 ft (180 cm) long and could propel a 3-ft (90 cm) arrow 1,200 ft (365 m). A skilled archer could fire up to ten arrows in a minute. Arrows with hardened steel heads could penetrate plate armor and iron mail from 300 ft (90 m).
> - The crossbow was 3 ft (90 cm) long and could propel a 1½-ft (45 cm) arrow 900 ft (270 m). A skilled archer could fire up to eight arrows in a minute.

▼ Battleaxes could only inflict injuries when used in close combat—and within range of the opponent's ax.

▲ Although it is easy to construct, a slingshot was seldom deadly unless in the hands of an expert.

letting go of one of the thongs and launching the stone toward the enemy.

The most important missile weapon was undoubtedly the bow, which was first used for hunting around 30,000 B.C. Several important technical developments have occurred during the course of its lengthy history. One of the most important of these was the introduction of the composite bow around 3000 B.C. The

MILITARY AND SECURITY

▲ *Missile weapons used in medieval warfare: the spear, bow, and crossbow.*

composite bow (made of a mixture of wood, animal sinew, and horn) continued to be a highly effective weapon until the Middle Ages, when it was used to devastating effect by the forces of Mongol conqueror Genghis Khan.

In the 12th century two further developments increased the potency of the bow as a weapon. The first was the rediscovery of the crossbow. The idea of placing a bow laterally across a wooden stock and fitting it with a mechanical winding mechanism and trigger was first explored by the Chinese in the sixth century B.C, but the design did not spread to Europe until 1,800 years later. The crossbow's winding system gave it enormous power, and its iron bolts were able to pierce even the strongest armor. However, the weapon did have its drawbacks, the most notable of which were a lack of accuracy and a limited range.

These problems were overcome by the Welsh longbow. The longbow was made from yew, a wood that possesses the qualities of a composite bow. Around 7 ft (2 m) in length, the longbow was extraordinarily powerful and extremely accurate. The use of the weapon by the English was a pivotal factor in their victory over the French at Agincourt in 1415.

THE COMPOSITE BOW

The effectiveness of the archer in battle was increased by the invention of the composite bow around 3000 B.C. This weapon had a long range and considerable penetration power. Composite bows were made by binding animal sinews to the front and horn to the rear of a curved wooden stave. The outer arms of the bow bent away from the archer, forming a reverse curve. The power of the bow relied on stringing the bow so that the layers of horn and sinew were bent back toward the archer (stretching the sinews and compressing the horn); this built up extra tension, which propelled the arrow forward with a considerable amount of force.

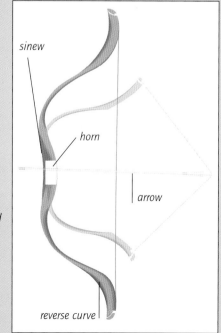

▶ *Composite bows were small yet powerful and ideal for use on horseback.*

WARFARE BEFORE THE CIVIL WAR

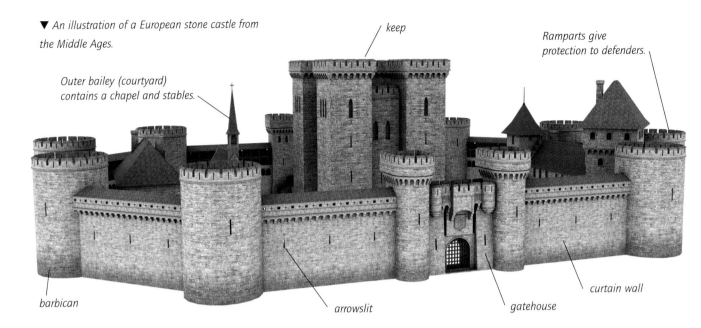

▼ An illustration of a European stone castle from the Middle Ages.

Outer bailey (courtyard) contains a chapel and stables.

keep

Ramparts give protection to defenders.

barbican

arrowslit

gatehouse

curtain wall

Fortifications and siege warfare

People began putting defensive walls around their homes and settlements tens of thousands of years ago—fortifications have been found at Jericho in the Middle East that date back to around 8000 B.C. Until 900 B.C. attacking armies lacked a reliable way of assaulting stone forts and surrounded them instead, forcing the occupants to come out fighting or surrender from starvation. Around that time, however, the Assyrians (from present-day Iraq) developed techniques to attack forts. They relied on machines such as the battering ram. Built on a six-wheeled wooden frame and covered by protective plates of leather, the Assyrian battering ram had a turret for observation and defensive fire.

The Assyrians were also prepared to go over the walls by using scaling ladders and by constructing ramps. Mantelets (large shields sometimes mounted on wheels) protected advancing soldiers from archers within the fort.

ATTACKING A CASTLE

When the Greek ruler Alexander the Great invaded Persia (now Iran) and India between 336 and 323 B.C., his engineer Diades invented a machine called the crow. This was a long, heavy bar suspended from a vertical frame that was used to knock down the upper parapets of a wall. Diades also invented another siege weapon called the telenon. It was a box large enough to hold a number of men that was slung from a boom suspended from a vertical frame. The basket could be raised or lowered and was used to place soldiers directly onto the enemy's walls.

▶ Attacking soldiers use a telenon to get over the castle ramparts.

MILITARY AND SECURITY

KEY COMPONENTS

Sambuca

The sambuca (siege ladder) was designed by the ancient Greek engineer Damis of Colophon. Soldiers climbed a ladder to enter the top compartment. Stones and rocks were loaded into boxes at the rear, raising the men at the front to the level of the battlements.

Fire-raiser

Fire-raisers were long wooden beams that carried cauldrons of lighted coals, sulfur, and pitch, which were used to start fires in the enemy's stockade. The wooden beam was hollow with a central iron tube and one iron-clad end to prevent it catching fire. A soldier pumped bellows at the other end to keep the fire alight.

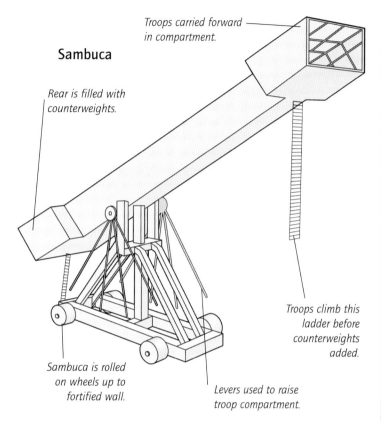

Sambuca

Troops carried forward in compartment.

Rear is filled with counterweights.

Troops climb this ladder before counterweights added.

Sambuca is rolled on wheels up to fortified wall.

Levers used to raise troop compartment.

Battering Ram

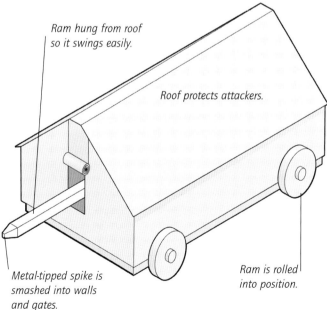

Ram hung from roof so it swings easily.

Roof protects attackers.

Metal-tipped spike is smashed into walls and gates.

Ram is rolled into position.

Battering ram

Battering rams were used to shake down a section of a wall. The one shown here was used in the fourth century B.C. The ram is metal plated at the battering end and housed in a so-called tortoise shell made of fire-resistant compressed seaweed and ox-hide. The ram was pushed forward on its wheels with considerable force and then pulled back by ropes and pulleys.

WARFARE BEFORE THE CIVIL WAR

▲ *A papyrus artwork shows an Egyptian war chariot. The archer on board can move around the battlefield at speed.*

around a fixed axle and carried two people. Ancient armies used chariots combined with infantry (foot soldiers), archers, and spearmen. With an archer on board, the chariot was a mobile platform of firepower. This was the style of warfare between 2500 and 900 B.C.

Chariots were more effective than cavalry because without stirrups cavalrymen could not

Mechanical artillery

Ancient artillery consisted of machines used for hurling heavy missiles. The invention of mechanical artillery is traditionally attributed to the tyrant King Dionysius the Elder (405–367 B.C.) of Syracuse, on the coast of the island of Sicily. Large stone-throwing catapults and machines (ballistas) capable of firing giant arrows a distance of 1,500 ft (450 m) were first used as early as 500–400 B.C., though. Dionysius made effective use of such machinery in his conquest of southern Italy between 379 and 367 B.C. The Macedonians produced lighter versions of the catapult and ballista that could be used on the battlefield.

Cavalry and chariots

The wheeled chariot first appeared in Mesopotamia (present-day Iraq) in 3000 B.C. By 1800 B.C., the light two-horse chariot was in use. It had spoked, hubbed wheels that turned

 SCIENTIFIC PRINCIPLES

Mechanical efficiency

This diagram shows how improvements over the years boosted the firepower of the ancient world's mechanical artillery. The degree to which the arms of artillery machines could be bent back toward the rear of the machine increased. This, in turn, increased the amount of twist in the rope springs to which they were attached, hence increasing the machine's firepower.

1 The arms of an early Greek arrow-firer could be moved 35° from a starting point of 20°.
2 The curved arms of the scorpio from c. 50 B.C. allowed them to be moved an extra 10°, twisting the rope spring even farther.
3 Around A.D. 100 the cheiroballistra had a metal frame, which allowed the angle of twist to be increased since the springs could be placed farther apart.

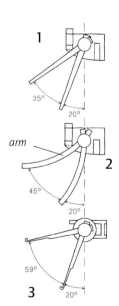

MILITARY AND SECURITY

KEY COMPONENTS

Onager catapult

The Roman catapult was also known as an onager, or "wild ass," because of its heavy recoil kick. The catapult is shown here just before firing. The trigger bar was normally struck by a hammer to ensure a clean release.

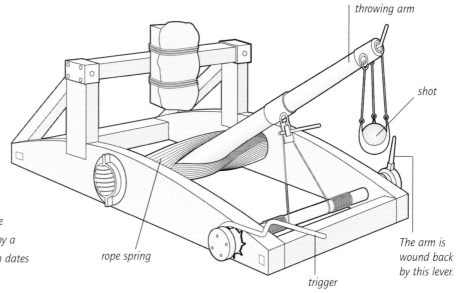

▶ The onager fired stone shot. The throwing arm was flung forward by a coiled spring of ropes. The weapon dates from around A.D. 350.

---

swing heavy weapons nor easily resist being pulled from their mounts. The most effective fighting force of the ancient world was the Roman infantry, organized in legions. However, in A.D. 378 horse-borne warriors from a tribe known as the Goths destroyed the legions of the Roman emperor Valens. Between 400 and 1300, cavalry then evolved into the dominant arm of battle in both Europe and the Middle East. This dominance was based on several technological factors.

Most crucial was the invention of the stirrup and related technologies like the saddle and bridle with bit, which allowed riders greater control over their horses. When combined with a saddle, the stirrup effectively welded rider to horse. This made it easier for a mounted soldier

BALLISTA

The Roman ballista shown here dates from around 25 B.C. It looked like a large crossbow and was a torsion weapon, using tightly twisted ropes to provide the propelling force. To load the ballista, a sliding trigger was pushed forward until it could hook onto the firing string. Then the trigger was wound back by a lever, pulling the string very tight. A ratchet prevented the string from flying forward until released.

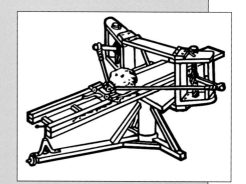

11

WARFARE BEFORE THE CIVIL WAR

▶ A halberd is the fusion of an ax and a long staff. It was an effective infantry weapon against cavalry attacks.

to charge into the enemy and to maintain a great deal of control over the horse while fighting. Stirrups and saddles probably originated in the Central Asian steppes between 300 B.C. and A.D 100. Stirrups were first used in Western Europe sometime between 500 and 1000.

Improvements in metalworking helped reinforce the dominance of cavalry and led to the development of heavy armor suits. Swords became longer, up to 40 inches (1 m) long, while by 1050 the spear was replaced by the lance—a long wooden shaft with a sharp metal spearhead—used for knocking knights off their horses.

Horse-borne troops continued to dominate warfare in Western Europe until the mid-14th century, when the emergence of the crossbow and the longbow tilted the balance of power in favor of the foot soldier. Another important factor in the cavalry's downfall was the reemergence of the pike in the 15th century. The pike was a wooden staff, typically about 12 ft (3.5 m) long, fitted with a sharp, piercing head. The weapon was held in battle, not thrown, and could unseat charging riders. Pikes were used by the Swiss infantry to inflict a number of heavy defeats on the knights of Burgundy and the Holy Roman Empire between 1460 and 1499.

ELEPHANTS

The use of elephants in warfare originated in India and spread to Greece and Carthage (now Tunisia in North Africa). Elephants inspired fear in troops and horses, but resourceful enemies learned how to make them stampede and become a danger to their own side. Anti-elephant measures included iron spikes chained to the ground to rip the animals' feet—a forerunner, perhaps, of antitank mines.

▲ A stone carving from Cambodia shows elephants marching into battle alongside foot soldiers.

The age of gunpowder

It was the invention of gunpowder that completed the demise of the knight. The origin of gunpowder is not really known. A formula for fireworks was provided in an 11th-century Chinese manuscript. This was a slow-burning substance, however, and was not the explosive of today. Both the Chinese and Indians used it for filling rockets. In China these rockets were used mainly to drive away evil demons and possibly also in wars against the Mongols in the 13th century. There is also some evidence that earlier, in the 12th century, the Chinese invented

MILITARY AND SECURITY

▶ The earliest record of gunpowder being used in Europe comes from German alchemist and monk Berthold Schwarz who is said to have discovered the explosive by accident somewhere between the years 1313 and 1353.

primitive grenades. These were essentially bamboo tubes filled with powder as well as stones, broken porcelain, and iron balls. More definite is that the Chinese were using bamboo guns in the 13th century, and that by the 14th century they were making guns made of metal.

Knowledge of China's black powder, as it was known, reached Europe, probably via the Mongols, in about the 1200s. Nearly another

SOCIETY AND INVENTIONS

The age of chivalry

The cost of a typical 16th-century knight's equipment was estimated to be $150,000 in today's prices. Such a high price ensured that all knights generally came from wealthy families. The barriers to achieving this rank were increased further because of the skill needed to fight as a knight. Training usually started in childhood and continued throughout life. The military importance of the knight was such that monarchs were prepared to make grants of land to knights in return for military service. This land was tended by peasants, and the knight lived off the surplus. This allowed him to purchase his equipment and to spend a lot of his time training, and showing off his skills at tournaments (right). Eventually, access to this warrior class was based solely on heredity. As a result, knights became part of an aristocracy that controlled Europe for the better part of a thousand years between the 8th and 19th centuries.

WARFARE BEFORE THE CIVIL WAR

200 years passed before gunpowder was used as an explosive propellant in European warfare. Two men are credited with being the first Europeans to describe in detail the making of gunpowder: English scientist Roger Bacon (1214–1294) and English scholar Albertus Magnus (c. 1200–1280). The first recorded Western use of gunpowder was by the English at the Battle of Crécy in 1346.

The first gunpowder cannons were essentially small metal pots packed with powder and then ignited to fire metal bolts. By the late 14th century enormous versions of these metal pots, called bombards, fired stones weighing as much as 300 lbs (140 kg). Developments in metalworking allowed the production of cannons with longer barrels, which were capable of greater range and power. Around 1450 stone shot was replaced by cast-iron balls. Because of their size and weight cannons were used primarily in siege warfare throughout most of the 15th century. By the 1490s, however, the French had developed light bronze-cast cannons. These were mounted on two-wheeled carriages pulled by horses and could be quickly brought onto the battlefield.

Handguns

The earliest handguns were developed at the same time as cannons. These were simple tubes of iron or brass with a hole in the top. Powder and ball were placed in the tube, the firer then held the gun with one hand while igniting the powder through the touchhole on top with a

▼ *Two gunners aim a cannon using clinometers, devices for measuring the angles of inclines.*

LIGHTER CANNONS

King Gustavus II (1594–1632) of Sweden made great strides in improving cannons. Before his time, cannons were too heavy to move once they were on the battlefield. Gustavus reduced the weight of the guns by making barrels shorter and thinner, but maintained their strength by casting them from a lighter mixture of copper and iron. The smaller cannons could be moved around the battlefield to assist infantry and cavalry attacks.

MILITARY AND SECURITY

THE MATCHLOCK MUSKET

One of the first handguns was the matchlock musket (right), invented by the Spanish military in the 1530s. Its long barrel made it powerful enough to penetrate all types of armor. It was an unwieldy weapon and had to be rested on a metal fork when fired. Although it was more accurate and had a greater range than any other handgun design of the time, it took time to reload once fired, taking 97 separate movements to fire. This resulted in a low rate of fire, perhaps one shot every two minutes. In spite of these drawbacks, the matchlock musket became the main firearm of European armies in the 17th century. The engineer-king, Gustavus II of Sweden simplified the loading process with a paper cartridge that contained gunpowder and a musket ball, doubling the rate of fire.

red-hot iron. As well as being very noisy, these small hand cannons were also inaccurate, and their barrels would become too hot to hold after only a few shots had been fired.

In the 15th century improvements were made in the ignition and accuracy of these firearms, and they became a class of weapon distinct from cannons. The touchhole was moved from the top to the side, and a small

SHRAPNEL

In 1784 British army officer Henry Shrapnel (1761–1842) invented the shrapnel shell. This was a canister filled with explosives and bullets. When fired from a cannon, it exploded in the air, spraying metal over a wide area. It was a deadly weapon against both infantry and cavalry.

pan was added to hold priming powder. A cord soaked in saltpeter (potassium nitrate, a flammable powder extracted from rocks), so that it continued to smolder after being lit, was used to ignite the weapon. This meant that the gunner did not have to stand close to a fire in order to use his weapon.

Ignition and firing systems were further improved with the development of the matchlock. This was an S-shaped mechanism attached to the side of the gun that held the burning cord and, when depressed by means of a trigger, ignited the powder. The barrel was placed within a wooden stock, and a curved wooden butt was added to the end of the barrel. This meant that the soldier firing the gun could hold the wooden butt piece against his shoulder while aiming and firing the weapon in the same way that a soldier today would fire a rifle. This gun was called a harquebus, meaning hooked tube. The harquebus was the standard infantry weapon until the middle of the 16th century.

At first pikemen defended musketeers while they reloaded, but by the end of the 17th century, the pike had finally been abandoned and all soldiers were armed with a musket. However, musket-armed infantryman were still vulnerable during the time spent reloading, pushing the shot into the barrel using a ramrod.

Two inventions made it possible for infantry soldiers to defend themselves. The first was the development of the bayonet, a dagger that attached to the musket, turning it into an effective pike. Second, the firepower of the infantryman was increased as the matchlock musket was replaced by the flintlock musket at the end of the 17th century. Flintlock muskets were quicker to load, cheaper than other mechanisms, and easier to repair since they had fewer small parts.

THE BAYONET

The advent of the musket did not result in its universal adoption by European armies. Its inaccuracy, slow rate of fire, and unreliability made the musket-armed infantryman vulnerable to attack from a host of traditional weapons such as the sword, ax, and pike. Until the development of the bayonet, essentially a knife that was attached to the barrels of guns, musket-armed infantry had to be supported by pikemen. The earliest bayonets were developed in Bayonne, France. The first bayonet was the plug bayonet. Devised in the 1640s, it was attached to the musket by inserting its handle into the gun's barrel. Once the bayonet was fixed, however, the musket could not be fired. In response to this it is believed that French military engineer Marquis de Vauban (1633–1707) constructed the socket bayonet. This was attached to the outside of the barrel and allowed the gun to be fired or used as a minipike.

▼ *Bayonet blades have a hollow in the side, preventing them from getting stuck in a victim.*

MILITARY AND SECURITY

KEY COMPONENTS

Matchlock

The matchlock firing mechanism was developed throughout the 15th and 16th centuries. It employed an S-shaped mechanism called a serpentine, which held a slow-burning saltpeter cord, or match. The serpentine was pivoted. Pushing the trigger lifted the bottom of the mechanism and brought the match down, bringing it into contact with the priming powder that fired the gun.

Flintlock

The matchlock was replaced by the flintlock in the 17th century. The flint was held in a vice known as a cock. When the trigger was pulled, the cock was propelled onto a piece of steel called the frizzen, which covered the priming powder. The blow not only produced sparks as flint hit steel, it also knocked the frizzen forward, uncovering the pan, and allowing the powder to be lit.

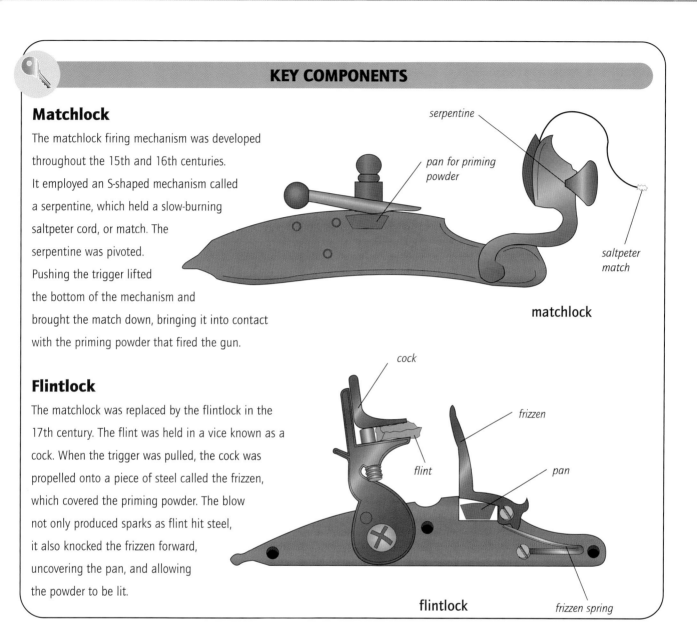

PERCUSSION CAP

In 1815 U.S. inventor Joshua Shaw devised the percussion cap. He enclosed the priming powders in a metal tube. The cap was detonated by a hammer on the musket, striking a pin within the tube. The percussion-cap musket was much more reliable than the flintlock musket and even fired during rain.

Despite this step forward in handguns, problems still remained. For example, it took a soldier a long time to ram a ball down the barrel of his gun. For this problem to be overcome, the handgunner would have to await the arrival of a new form of ammunition—the bullet—in the late 19th century.

CIVIL WAR TO THE NUCLEAR AGE

Initially, the Industrial Revolution did not make a major difference in the way that armies were armed or fought battles. By the American Civil War (1861–1865), however, it was clear that inventions from industry were changing the nature of warfare forever.

The first weapon to be changed dramatically was the infantry (foot-soldier) rifle. In Prussia (now part of Germany) Johann Dreyse (1787–1867) invented the first modern breech-loading rifle, the "needle gun," which entered military service in 1848. However, the design of the weapon was a closely guarded secret, and at the start of the Civil War most soldiers were still

▼ Workers mass produce revolvers at the Springfield Armory in Massachusetts in the 1880s.

▲ The huge mortars used during the Siege of Yorktown in 1862 weighed 10 tons (9 metric tons) each.

using the older muzzle-loading rifles. It was during this war that the breech-loading rifle became widespread. Tyler Henry, an American gunsmith, introduced the Henry repeating rifle, while Christopher Spencer invented a short-barreled rifle, or carbine. In both of these breech-loading rifles ammunition was stored in an attached compartment, called a magazine, which allowed more rapid firing. A modified version of the Henry rifle, called the Winchester, was widely used in the Wild-West era.

By the 1880s the Springfield Armory in the United States, Enfield Arsenal in Britain, and Mauser factory in Germany had all produced fairly similar robust infantry rifles. These were magazine-fed, breech-loading rifles, generally of

MILITARY AND SECURITY

TYPES OF RIFLE

1 Muzzle-loading rifle

To load this type of rifle, powder (usually black gunpowder) and shot (usually a single lead ball) were inserted down the end of the barrel (pictured right) and rammed past the spiral grooves inside with a metal rod. This procedure is awkward and slow; even well-trained soldiers could only manage about three shots per minute.

▶ Shotgun cartridges are loaded at the base of the barrel, and the hinged stock is locked back into place before firing.

2 Breech-loading (or bolt-action) rifle

In a breech-loading rifle, the powder and ball are inserted as a unit into the breech of the rifle, which can be opened by sliding back a steel cylinder, called the bolt. When in a forward position, the bolt seals the breech and allows firing. Breech-loading rifles can be fired much more rapidly than their muzzle-loading predecessors.

3 Self-loading rifle

A self-loading rifle works by using some of the exhaust gases from firing to drive a reloading mechanism. Usually there is a small hole drilled into the barrel of the rifle through which gas escapes into a metal cylinder that runs back toward the breech. The bolt is driven back either by the gas itself or by a gas-operated piston. The spent cartridge is ejected and a new one loaded from the magazine as the spring-loaded bolt returns to a forward position.

4 Automatic rifle

The automatic rifle works by the same mechanism as the self-loading rifle, but the trigger does not need to be released between shots. This allows continuous firing. The main problem with automatic rifles is in slowing the rate of fire to a controllable level. This is usually achieved either by using a very heavy bolt that takes longer to travel backward and forward or by using a mechanism to slow the bolt's travel.

CIVIL WAR TO THE NUCLEAR AGE

Bolt-action rifle

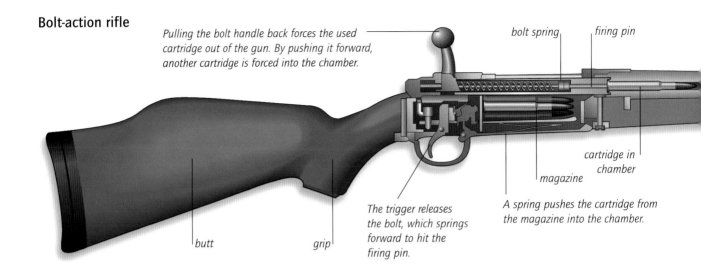

Pulling the bolt handle back forces the used cartridge out of the gun. By pushing it forward, another cartridge is forced into the chamber.

bolt spring

firing pin

cartridge in chamber

magazine

A spring pushes the cartridge from the magazine into the chamber.

The trigger releases the bolt, which springs forward to hit the firing pin.

butt

grip

about 0.3-inch (7.62 mm) caliber (the diameter of the inside of the gun barrel) and capable of ten aimed shots per minute. Reliable and effective, these weapons were the standard infantry rifles throughout both world wars.

During the 1930s attempts were made to further improve the firing rate of infantry rifles. U.S. engineer John Garand developed a self-loading rifle. His design was adopted by the U.S. Army in 1936, giving U.S. soldiers the most modern infantry weapon at the time.

During World War II (1939–1945) the Germans developed the MP-44, the first assault rifle—an accurate weapon capable of single-shot and automatic fire. This design was copied by the Russian engineer Mikhail Kalashnikov (born 1919) to produce the AK-47, probably the most widely used infantry rifle in the world today.

Improving artillery

Artillery pieces did not change as quickly as the rifle during the 19th century, but as infantry muskets began to be rifled, so gun manufacturers tried to rifle artillery pieces. The

▲ A World War I rifle could hold ten rounds, or cartridges, in its magazines and could hit targets up to 2,000 ft (600 m) away.

GUN METALS

The technique of blasting molten pig iron with air to produce high-quality steel was devised by British metallurgist Henry Bessemer (1813–1898) and changed the manufacture of guns dramatically. Using steel, guns could be made much stronger, lighter, and more powerful. Manufacturers like Alfred Krupp (1812–1887) in Germany used tougher steels to produce breech-loading, rifled artillery pieces that were far superior to the older designs.

▲ Guns made with the super-strong steel developed by Henry Bessemer could fire heavy high-explosive artillery shells over several miles.

MILITARY AND SECURITY

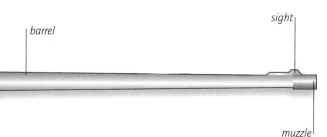

barrel

sight

muzzle

SCIENTIFIC PRINCIPLES

The benefits of rifling

Most firearms made before the Civil War were of the smoothbore type—the inside of the barrel was simply a smooth cylinder. These weapons fired loose-fitting bullets that could be pushed easily from muzzle to breech. This meant that the bullet tended to wobble as it traveled up the barrel after the gun was fired, reducing accuracy. The advantages of rifled weapons, which have a series of coiling grooves inside their barrels, had long been known. The grooves make the bullet spin as it leaves the barrel, leading to a more stable flight and greater accuracy. For this effect to work, however, the bullet must fit tightly inside the barrel, making it difficult to ram from muzzle to breech and reducing the rate of fire. A solution to this problem was found in 1841 by Etienne Minié. He developed a cone-shaped bullet that was smaller than the diameter of the barrel. This allowed the bullet to be dropped down the barrel. When fired, however, the gases produced in the barrel caused the bullet to expand, fitting it into the grooves of the rifle, and this ensured that accuracy was maintained. With the invention of breech-loading mechanisms, rifled barrels became standard in both infantry weapons and artillery.

▼ *The rifled barrel of this train-transported WWI artillery gun caused shells to spin so they flew straight to a distant target.*

CIVIL WAR TO THE NUCLEAR AGE

problems associated with rifling heavy weapons were considerable, and before much progress could be made, a method of manufacturing stronger gun barrels had to be found. In 1854 William Armstrong (1810-1900), a British engineer, made the first reliable rifled artillery barrels. Constructed from a series of forged iron tubes, Armstrong's barrels were far more accurate than the old smoothbore types and were quickly adopted by the British Army.

The real breakthrough for artillery came in 1897, when the French Schneider company produced a new and radical 3-inch (75 mm) gun. This weapon not only had a rifled barrel and breech loading but also a liquid-filled recoil system, which returned the barrel to its original position after firing. Because it did not have to be moved between shots, the "75" could fire six aimed shells a minute—far more than any other artillery piece at that time. In fact, although artillery pieces would be developed

◄ A molecule of nitroglycerin has nitrogen atoms (blue) loosely bonded to oxygen atoms (red). During an explosion these bonds break releasing a lot of energy very quickly.

ALFRED NOBEL

Alfred Nobel (right) was a brilliant and determined Swedish engineer and chemist. He built up a large manufacturing business and, fascinated with the properties of explosives, he began intensive and dangerous research to find better explosives than gunpowder. Experimentation with guncotton and nitroglycerine led him to discover several new explosives that were stable, safe, and very powerful. Among these were dynamite and blasting gelatin, which are used in industry, and ballastite, which became the basis for most modern ammunition. Treated with suspicion and even disgust by many of his countrymen for his research into deadly explosives, Nobel left much of his considerable fortune to a foundation that established five open competitions in areas of human endeavor. One of the most famous is the Nobel Peace Prize.

MILITARY AND SECURITY

▲ Much of the work at the Nobel dynamite factory was done by women, who mixed the nitroglycerin with earth and sawdust to make it more stable.

armies. However, Swedish chemist Alfred Nobel (1833-1896) made a breakthrough in 1875 by combining guncotton with nitroglycerine to make dynamite, a substance that is stable until detonated and burns without smoke. From dynamite, which was used mainly in mining and further during the 20th century, with higher rates of fire, longer ranges, and heavier explosive charges, they all followed the pattern of the French 75.

Hand in hand with the inventions of new rifles and guns came the development of new, more powerful explosives. Gunpowder had been used for centuries, but its explosive force was relatively limited. In 1846 Austrian chemist Christian Schonbein (1799-1868) produced guncotton by adding nitric acid to cotton. While guncotton was far more powerful than gunpowder, it was also highly unstable and dangerous and was not adopted by many

BIGGER BANGS

While ballastite and cordite became the main explosives used for firing guns, the explosive most widely used to fill shells was picric acid, first tested by the French in 1886. A safer, more stable alternative to picric acid, TNT (trinitrotoluene), was discovered in Germany in 1902 and became the standard explosive for shells. TNT is much more powerful than gunpowder, and long-range artillery bombardments of immense destructiveness were a grim feature of both world wars. During WWII the British company ICI developed the first plastic explosives. These were safe to handle, easy to mold into any shape, and difficult to detect. First used by British commandos and resistance groups in WWII, plastic explosives have since been used around the world by terrorist groups.

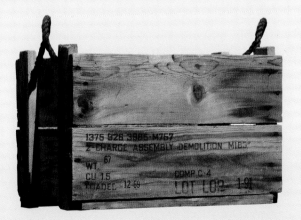

▲ Crates of plastic explosives like this are used to demolish large buildings in seconds.

23

CIVIL WAR TO THE NUCLEAR AGE

FLAME WEAPONS

Flamethrowers were developed by the German Army in the early 1900s and were used with surprise effect against Allied troops in 1915. The original flamethrower was a steel tube that squirted burning oil using compressed air. The British soon developed a similar device, but all flamethrowers in WWI were clumsy and short ranged. The development of napalm (jellied gasoline), which carried farther than ordinary gasoline, burned with intense heat, and clung to whatever it touched, made WWII flamethrowers more lethal. The British Crocodile tank, which towed an armored trailer filled with napalm, could fire 100 shots to a range of 260 ft (80 m). During the Vietnam War napalm bombs delivered by aircraft cleared large areas of land.

▼ A bomb test shows the destructive power of napalm bombs. The flames burn at 1,500°F (800° C).

industry, Nobel developed ballastite, a military explosive. English chemists Frederic Abel and James Drewer, who were working along similar lines, produced cordite, another stable, smokeless explosive.

The machine gun

Modern artillery was one of the dominant weapons in the world wars; the machine gun was the other. The idea of a repeating gun had been around for a long time, but it was only in the Civil War that practical designs were used in battle. U.S. inventor Richard Gatling (1818–1903) built the first reliable machine gun in 1862. It was hand-driven and consisted of ten barrels rotating around a central crank. During a full rotation each barrel was loaded and fired, and the spent cartridge was ejected. The French also developed a hand-driven machine gun, the Mitrailleuse, which fired 125 rounds per minute, but it was not used very effectively in the Franco-German War of 1870.

With the development of Nobel's more powerful smokeless explosives, a U.S. inventor, Hiram Maxim (1840–1916), developed the first fully automatic machine gun in 1884. Instead of being hand-cranked, the Maxim gun used the gas and recoil produced by firing to load the

MILITARY AND SECURITY

next round and could fire 500 rounds per minute. The Maxim gun was copied by most European armies and was used to deadly effect in World War I (1914–1918). To prevent the barrel from overheating during sustained firing, the Maxim gun was cooled with water, which made it heavy and difficult to transport on the battlefield. This led other inventors, notably Isaac Newton Lewis, to develop lighter air-cooled machine guns. The Lewis gun, invented in 1912, was adopted by the Allied armies in WWI and gave soldiers a mobile machine gun that revolutionized infantry tactics.

Grenades and mortars

Another simple infantry weapon that came into its own during WWI was the grenade, a

▲ Hiram Maxim shows off his machine gun in the 1880s. Maxim also invented the gun silencer and is credited with the development of the mousetrap.

KEY COMPONENTS

Machine gun

A general purpose machine gun is a weapon capable of performing a wide variety of roles, such as providing supporting fire or shooting down aircraft, yet light enough to be transported by infantry. The MG34 developed by the German Army in 1934, was the first weapon of this type, and its design was copied in the British Army's Bren gun (shown here). The gun can be mounted either on a bipod (two-legged) stand or on a tripod (three-legged) stand for shooting up into the sky. Bullets are clipped onto a magazine, which is connected into the breech. A gas-operated reloading piston ejects the used cartridge and loads the next bullet automatically. The gun is capable of firing 500 rounds per minute but is usually used in short bursts to prevent the barrel from overheating. A flash hider at the end of the barrel makes it more difficult for the enemy to locate the gun.

Bren gun

25

CIVIL WAR TO THE NUCLEAR AGE

hand-held explosive device that could be hurled at the enemy, detonating after five or seven seconds. Grenades provided soldiers with "pocket artillery," which was of great use in trench warfare.

Mortars were also developed to help soldiers in the trenches. Mortars (short-barreled, high-angled guns for dropping shells onto the enemy) had been in use for many years, but the first modern mortar, called the Minenwerfer, was designed by the Germans in 1908. The British and French soon produced their own versions, the Stokes and Brandt mortars. The

▲ This early 18th-century mortar, known as a "partridge," fired one large shell plus a dozen smaller grenades.

SOCIETY AND INVENTIONS

The cost of war

All of the developments in weaponry during the 19th and 20th centuries resulted in armies possessing vastly increased killing power. Riflemen could fire at least ten rounds a minute with great accuracy. The machine gun, described as the "concentrated essence of infantry," could deliver very high volumes of fire—advancing infantry were literally mown down. Artillery pieces firing explosives and gas killed thousands of people. On July 1, 1916, the British Army suffered 60,000 casualties in one day. Over the course of WWII, 32 million people died, as warfare became more mechanized and weapons more powerful. In the 20th century, wars regularly killed several percent of a country's entire population.

▼ A war cemetery in Normandy, France, marks the graves of just some of the American soldiers who died on D-Day, June 6, 1945.

MILITARY AND SECURITY

▶ Grenades are primed by pulling out a pin. The thrower releases the lever, which sets off a short time fuse.

German 3.2-inch (81 mm) mortar of 1939 was an extremely effective design consisting of a simple tube mounted on adjustable legs and resting on a base plate. The operator dropped the bomb down the tube to fire the mortar. Although refinements have been made, all modern mortars follow this basic design.

Chemical weapons

One of the most terrible weapons developed during WWI was poison gas. The Hague Convention of 1899 condemned the use of gases in warfare, but the Germans, French, and British all continued work on producing them. German chemist Fritz Haber (1868–1934) developed poisonous chlorine and phosgene gases for military use. The French had used tear gas against the Germans in August 1914, but it was only when the Germans used poisonous chlorine gas near Ypres in Belgium on April 22, 1915, that gas had a major impact on the war.

MINE WARFARE

The first mines were large bombs placed under an enemy position by miners, who dug tunnels under the battlefield. In response to the threat of tanks, inventors in the 1930s developed modern mines. These are buried containers filled with explosives and detonated by the pressure of a tank or person above them. Mines are easy to make and lay but difficult to detect and remove. Mines had a major effect, slowing or stopping enemy advances in several of the major tank battles of WWII.

▶ Mines are hidden weapons, and they are difficult to locate and make safe after a war has ended (main image). Six million mines were laid in Cambodia during the 1970s and 1980s, and 40,000 civilians have been killed or injured by them (inset image).

27

CIVIL WAR TO THE NUCLEAR AGE

Without gas masks, many of the French soldiers panicked, leaving a gap in the line, but the Germans were not able to exploit their success. Initially, poisonous gases were released from cylinders and was spread by the wind, but soon shells were filled with gas and fired onto enemy positions. Gas was widely used during WWI and killed over 90,000 soldiers.

Armored warfare

Armies were quick to realize the advantages of the internal-combustion engine for transporting weapons. By 1910 automobiles were being used as troop transporters and mobile machine-gun carriers. The idea of a tracked, armored vehicle had been thought of by the great Italian artist Leonardo da Vinci (1452–1519), but it was not until 1915 that British engineers Ernest Swinton and Walter Tritton built the first practical vehicle. Codenamed "Water Tanks for Russia" for secrecy, the name *tank* has stuck to this day.

While the tanks used in WWI were heavy and slow, they could crush barbed wire and help infantry attack enemy trenches. In WWII fast, well-armed tanks (and dive-bombing aircraft) were used by Germans forces in a new, fast-moving form of warfare called *Blitzkrieg*.

By 1945 warfare had become very technical and sophisticated, and most of the major developments in infantry weapons, artillery, and tanks had been achieved. However, with the invention of the atomic bomb by the Allies, and guided rockets by the Germans, warfare entered an entirely new and more destructive era.

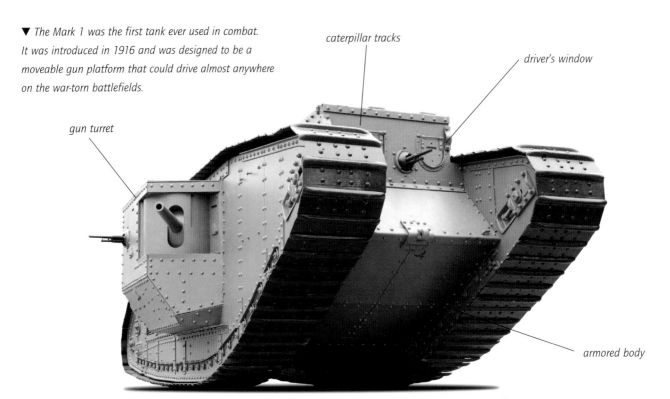

▼ The Mark 1 was the first tank ever used in combat. It was introduced in 1916 and was designed to be a moveable gun platform that could drive almost anywhere on the war-torn battlefields.

gun turret

caterpillar tracks

driver's window

armored body

MILITARY AND SECURITY

▲ British children wear gas masks during WWII as a defense against chemical-weapon air raids. No such attacks were made on the U.K. during the war.

FACTS AND FIGURES

- The British Mark 1 tank was armed with six-pounder guns and machine guns. It weighed 31 tons (28 metric tons) and had a top speed of 4 mph (6 km/h).
- The Russian T-34 tank was produced in vast numbers during WWII. Simple and robust, it weighed 39 tons (36 metric tons) and carried a 3.3-inch (85 mm) gun.
- The U.S. Sherman tank was the main tank used by the Allied armies during WWII. It weighed 36 tons (33 metric tons) and was armed with a 3-inch (75 mm) gun.
- The German King Tiger tank was the most powerful armored fighting vehicle of WWII. Armed with a 3.5-inch (88 mm) antitank gun, it had 9-inch (230 mm) thick steel armor and weighed 76 tons (69 metric tons).

THE HOLOCAUST

While chemical weapons were not used in battle during WWII, Nazi Germany employed poisonous gases in Adolf Hitler's Final Solution—the extermination of the "undesirable people." Throughout the German-occupied areas of Europe, Jews, Roma (gypsies), homosexuals, and disabled people were systematically killed. Some became slaves in German factories until they perished from exhaustion or starvation, and many more were sent to death camps, such as Auschwitz in Poland. They were killed in large groups inside gas chambers. Called the Holocaust, this persecution claimed the lives of at least 11 million people.

▶ The gate at Auschwitz welcomed inmates with the words "Work makes you free." However, Auschwitz was not a work camp, but was designed to exterminate more than 1,000 people every day.

MODERN WARFARE

World War II was brought to an end using a new weapon of terrifying power, the nuclear bomb. Since then military technologies have been developed to strike any point on Earth with devastating force.

In early August 1945 atomic bombs were dropped on the Japanese cities of Hiroshima and Nagasaki, causing death and destruction on an immense scale and leading to Japan's surrender and the end of World War II (1939–1945).

The groundwork for the development of nuclear weapons was laid by British physicist Ernest Rutherford (1871–1937), who discovered that atoms, the building blocks of all matter,

H-BOMB WARHEAD

This is one of several designs of a thermonuclear bomb, which uses heat and nuclear reactions to create an explosion. The fission device is a small nuclear bomb that explodes creating huge temperatures. This causes the lithium deuteride to release radioactive hydrogen atoms, which fuse with each other releasing an immense amount of energy. This second explosion causes a third in the uranium jacket.

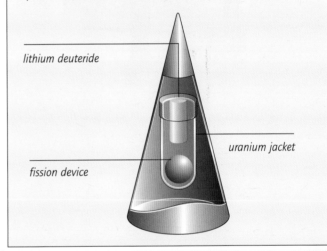

▼ The first hydrogen bomb is tested on the Pacific island of Enewetak in 1952 by U.S. researchers. The explosion produced the mushroom cloud typical of nuclear bombs.

MILITARY AND SECURITY

were made up of several smaller particles. Rutherford also demonstrated that the atoms of some elements could be "split" to form other elements. The use of these discoveries did not become apparent until 1939, when two German chemists, Otto Hahn and Fritz Strassmann, succeeded in splitting the uranium atom, a process that released a vast amount of energy, as predicted by German physicist Albert Einstein (1879–1955). If a reaction of this type could be performed on a larger scale, it would produce an immensely powerful explosion. Research into nuclear weapons was never given a high priority in Nazi Germany, but the U.S. Manhattan Project, led by J. Robert Oppenheimer and involving many of the pioneers of nuclear physics from both the U.S. and Europe, produced the first atomic bomb in July 1945. Initially, only the United States possessed the atomic bomb.

However, in 1949 the Soviet Union tested its own atomic bomb. Britain, France, and China were also quick to develop weapons of this type.

The hydrogen bomb

In 1941 two U.S. physicists, Enrico Fermi (1901–1954) and Edward Teller (born 1908), working on the Manhattan Project saw that it would be possible to use the energy from an atomic bomb to trigger an even more powerful nuclear reaction in which atoms of deuterium, a rare form of hydrogen, are fused together.

In 1949, when the Soviet Union developed an atomic bomb, the United States decided to build one of these "hydrogen bombs." When detonated, a single hydrogen bomb could produce an explosion equivalent to millions of tons of TNT. By August 1953 the Russians had their own hydrogen bomb.

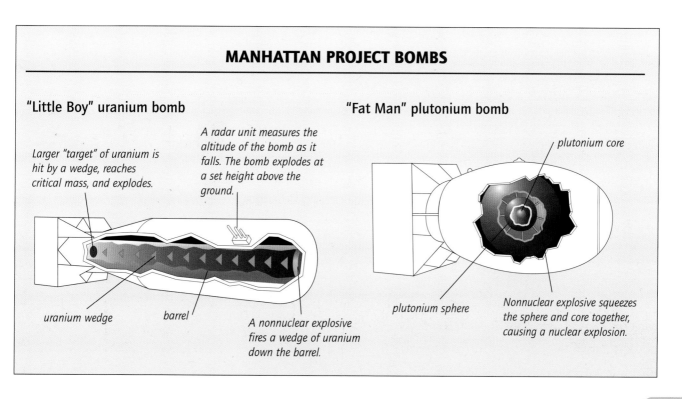

MANHATTAN PROJECT BOMBS

"Little Boy" uranium bomb

Larger "target" of uranium is hit by a wedge, reaches critical mass, and explodes.

A radar unit measures the altitude of the bomb as it falls. The bomb explodes at a set height above the ground.

uranium wedge

barrel

A nonnuclear explosive fires a wedge of uranium down the barrel.

"Fat Man" plutonium bomb

plutonium core

plutonium sphere

Nonnuclear explosive squeezes the sphere and core together, causing a nuclear explosion.

MODERN WARFARE

▲ When a message was typed on the Enigma machine's keyboard, each letter was converted into another, shown by lights on the upper set of letters. The code was controlled by a series of adjustable cogs inside the machine, which could be set into more than 17,000 different combinations.

Computerized warfare

The first computer used in war was developed in the 1930s by Polish scientists who used it to break German military Enigma codes. Influenced by the success of this machine, a team of British scientists and engineers led by Alan Turing (1912–1954) built a far more powerful computer in 1943. Called Colossus, this computer was highly successful at decoding German messages during WWII. After the war, a number of large military computers were constructed. Although there were great improvements in processing power and memory, these machines were cumbersome and unreliable. The miniaturization of computers in the 1960s and 1970s allowed their use in a wide variety of military roles where rapid and accurate calculations are needed. Computers are now vital components of guided missiles, radar, artillery targeting, and most importantly battlefield communications.

Communications

In both World War I (1914–1918) and WWII communications were conducted by the telegraph, the telephone, and the radio. During

TACTICAL NUCLEAR WEAPONS

Although the power of the first nuclear weapons was almost limitless, their usefulness on the battlefield was limited by the fact that they were so large and heavy. During the 1950s, however, nuclear weapons that were small enough to be fired from artillery pieces or carried by fighter aircraft were developed by both the United States and the Soviet Union. These weapons, designed to be used against masses of enemy troops or tanks rather than large civilian targets, are called tactical nuclear weapons.

▲ The B-1 Lancer bomber was designed for the U.S. Air Force to carry nuclear weapons deep into enemy territory.

MILITARY AND SECURITY

SOCIETY AND INVENTIONS

Nuclear weapons and society

The existence of nuclear weapons created a sense of unease in the societies of the industrialized West. What would happen if both the Soviet Union and the United States launched their entire nuclear arsenals? Would anyone survive? There were fears that a nuclear war might be started by a madman or by a tragic accident caused by human or computer error. Within Europe fears about nuclear weapons were mixed with a certain suspicion that the United States was planning to fight a nuclear war in Europe. These fears were increased in 1977, when the United States was debating whether or not to deploy the neutron bomb. This weapon was designed to kill through increased levels of radiation while reducing blast and fallout created by an explosion. In essence, the neutron bomb killed people without damaging property. Public hostility to this weapon was an important factor in it not being deployed. In the 1980s fear of nuclear war increased as scientists speculated about the terrible environmental effects of a nuclear war. One possibility is that huge clouds of dust would be blasted into the atmosphere, blocking out the Sun and causing a long nuclear winter that would kill most of the animal and plant life around the world.

STAR WARS

Efforts were made by both the United States and the Soviet Union to develop an antimissile defense capable of stopping nuclear attacks. In 1983 U.S. President Ronald Reagan announced plans to build a space-based antimissile defense system called the Strategic Defense Initiative (SDI), or "Star Wars." This envisaged the use of laser-armed satellites to destroy incoming ICBMs before they had a chance to release their warheads. The main problem with this proposal was the computing power needed to manage such a battle in space. It was well beyond the capabilities of existing or even proposed future developments. It was also estimated that Star Wars would have cost about one trillion dollars!

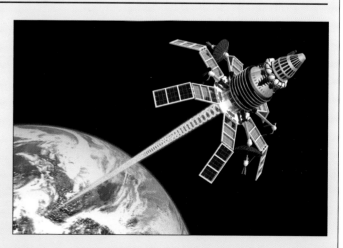

▲ Anti-missile satellites would have given the United States a critical advantage over its enemies. However, they were never built.

MODERN WARFARE

MILITARY SATELLITES

The military potential of satellites was fully appreciated by the U.S. as early as 1946, when the U.S. Navy discussed the possibility of using satellites to control guided missiles. However, the deployment of satellites had to wait until the mid-1950s, when more advanced rockets and electronics became available. The first U.S. military satellite, Samos II, was launched in January 1961 and provided photographs of Soviet nuclear missile sites. Throughout the 1960s and 1970s an increasing number of military "spy" satellites were launched by the United States and the Soviet Union. The quality of pictures taken from these satellites was so good that it was possible to see even minute details. Modern military satellites are capable of sending back instantaneous live-action video images.

Military satellites are also useful for navigation. The U.S. Global Positioning System (GPS) consists of a network of satellites that continuously relay the time and their locations back to Earth. Fighting vehicles, missiles, and combat troops are often equipped with small computers that can use this data to calculate their own coordinates to within a few feet.

▲ A satellite image shows up the roads, field boundaries, and even the snow surrounding a town in Central Asia.

battles both sides could often listen in on each other's communications, which were limited to short transmissions. Commanders might only have a general idea of where their forces were and how the battle was progressing.

Small, powerful computers and digital transmissions revolutionized communications in war. The volume of information that could be transmitted, received, and processed was dramatically increased. The location and status of every unit could be relayed by satellite to commanders hundreds of miles away.

Modern militaries rely heavily on communications and so blocking the enemy's communications is a useful weapon. This is called electronic warfare and involves

▼ Early warning radars contained within domes scan the sky in all directions for large flying objects.

MILITARY AND SECURITY

jamming the enemy's radar and radio links through the transmission of electronic signals. Also of importance is the use of surveillance systems to monitor all radio bands to collect intelligence on enemy communications and the location of forces. To counter these technologies, modern military telecommunication systems, such as the British army's Ptarmigan system, send computer-encrypted messages that are difficult for the enemy to decode and often change the radio frequency at which they transmit to avoid jamming.

Guided missiles

By the end of WWII the Germans had made significant advances in the development of guided missiles. The V-2 rocket was able to reach targets 200 miles (320 km) away, and several surface-to-air missiles, designed to shoot down enemy aircraft, had been tested. Both the United States and the Soviet Union made use of the German technology to build guided missiles of their own after the war. While steady improvements were made to the design and efficiency of rocket engines, it was the advances

---

KEY COMPONENTS

Cruise missiles

Cruise missiles are essentially uncrewed high-explosive aircraft. Their engines, which are often air-breathing jets rather than true rockets, burn for a prolonged period (often several hours), and they have short wings that allow them to fly. The German V-1 was the first "cruise missile." It had a speed of 350 mph (560 km/h), a range of 160 miles (260 km), and flew at 2,000–3,000 ft (650–1,000 m). The V-1 had a small jet engine, which took in air at the front, mixed it with fuel, and ignited the mixture. The resulting hot exhaust gases escaped from the rear, propelling the missile forward. The missile was steered using elevators and a rudder at the rear. Guidance was by autopilot—a compass linked to a gyroscope that was preset on the V-1's target before launch. The V-1 flew a preset distance and was then was put into a steep dive by lowering flaps on the tail section; the engine then cut out. The sudden silence gave a

terrifying warning to people on the ground that a so-called doodlebug was about to hit. More than 20,000 of these missiles were produced by Nazi Germany during WWII—most of them were made by slave workers.

The United States experimented with cruise missiles in the 1950s but found it impossible to develop an accurate guidance system. By the 1970s the miniaturization of computers and electronics allowed the creation of new and more accurate systems of navigation, which were incorporated in modern cruise missiles, such as the U.S. Tomahawk (above).

MODERN WARFARE

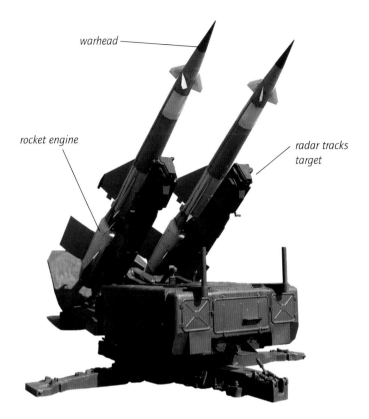

warhead
rocket engine
radar tracks target

▲ Anti-aircraft missiles are targeted by radar. When an aircraft is detected, the rocket launcher swivels toward it and raises the missiles to the right angle before firing.

in electronics, sensors, and computers that allowed guided missiles to become some of the dominant weapons on the modern battlefield.

Surface-to-air missiles

In the United States attention focused initially on developing a surface-to-air missile (SAM) capable of destroying enemy aircraft. By 1953 the U.S. had developed its first SAM, the Nike Ajax air defense missile. The missile was large, cumbersome, and could only be fired from fixed installations. A powerful boost motor accelerated the missile to speeds faster than sound, and a smaller sustainer rocket powered the missile over the rest of its flight. Once

---

FACTS AND FIGURES

ICBMs

U.S. Minuteman III: fixed land-based; 3 MIRVs; range 8,080 miles (13,000 km).

USSR's SS18: fixed land-based; 10 MIRVs; range 6,835 miles (11,000 km).

USSR's SS20: mobile land-based (below); 3 MIRVs; range 2,485 miles (4,000 km).

U.S. Pershing II: mobile land-based; 1 reentry vehicle; range 2,070 miles (3,300 km).

U.S. Trident I: submarine-launched; 8 MIRVs; range 4,350 miles (7,000 km).

MIRV stands for Multiple Independently Targeted Reentry Vehicle.

---

airborne, radar beams tracked both the missile and the target, sending information to a computer, which guided the missile onto a collision course. One year later the Soviet Union deployed its own SAM, the SA-1, which was guided by a radar located inside the missile itself—a more versatile system than that used in Nike Ajax. In the following decades both the United States and the Soviet Union developed SAMs that were smaller, more agile, and able to engage targets at greater distances.

MILITARY AND SECURITY

Intercontinental ballistic missiles

Until the 1950s the United States and the Soviet Union both relied on large strategic bombers to attack each other's homelands in the event of a war. However, the development of SAMs reduced the chances of bombers penetrating the enemy's missile defense systems. In response to this dilemma both countries developed intercontinental ballistic missiles (ICBMs) armed with nuclear warheads. These are multistage rockets similar to those used to put satellites into orbit. From there the warhead follows an arc, returning to Earth thousands of miles from the launch site. Initially, these missiles relied on inertial guidance, using gyroscopes to calculate their position. However, later ICBMs incorporated stellar navigation

▼ A Minuteman II ICBM is hidden in a deep underground silo. The missile has three rocket stages that power it into space and over a target in three minutes.

KEY COMPONENTS

Trident II missile

This model of ICBM is carried on U.S. and British nuclear submarines. They are launched under water and can carry up to 12 warheads more than 4,100 miles (6,600 km).

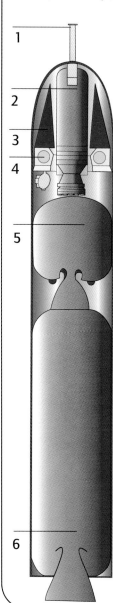

1. The aerospike reduces drag as the missile flies through air.
2. The third-stage rocket motor positions the warheads above their target while in space.
3. The missile can carry 12 MIRV warheads.
4. The equipment section contains the guidance and targeting systems.
5. The second stage rocket motor provides an extra boost to get the missile into space.
6. The first stage motor is used to launch the missile. Like the other stages, this motor is powered by solid rocket fuel.

MODERN WARFARE

SCIENTIFIC PRINCIPLES

Missile guidance

How does a missile know where to go when it is fired?

1 Radio control The missile is controlled by an operator sending commands by radio waves.

2 Infrared sensor An infrared sensor in the nose of the missile picks up heat emissions from the target. Infrared guided missiles are most useful against jet aircraft, which emit a lot of heat.

3 Terrain contour mapping (TERCOM) An on-board radar maps the terrain over which the missile is flying. A computer compares this data to a map stored in its memory.

4 Radar command A radar tracks the target while another tracks the missile. A computer compares the two sets of data and steers the missile toward the target using a radio link.

5 Active radar homing A radar in the nose of the missile detects the target, while an on-board computer steers the missile.

6 Wire guidance The missile unwinds a thin wire behind it as it flies. The operator tracks both missile and target, sending steering commands down the wire to the missile.

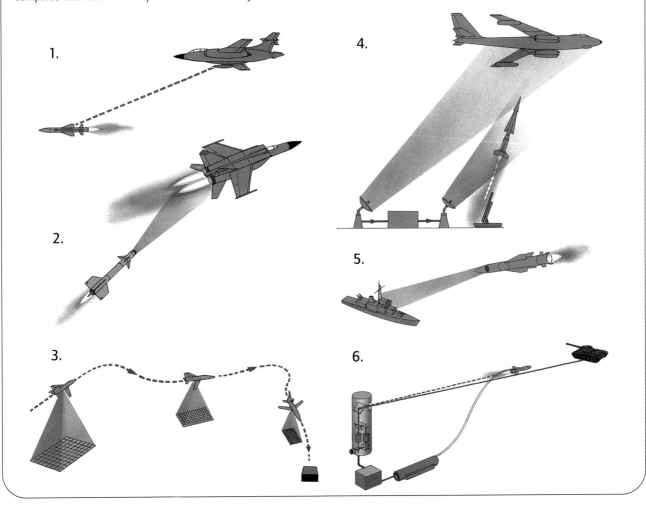

MILITARY AND SECURITY

(checking the position by reference to the stars) and radar to improve accuracy.

Throughout the 1950s and 1960s the United States and the Soviet Union developed ICBMs with ever greater ranges and more powerful warheads. The Soviet SS-9 was the first operational ICBM to incorporate multiple warheads, or Multiple Reentry Vehicles (MRVs). The missile carried several smaller warheads that separated from one another during the middle part of the missile's flight. This technology was

CHEMICAL AND BIOLOGICAL WARFARE

Poisonous gases were used as weapons in WWI. During the 1930s, the Germans developed far more deadly chemical agents. Among these were two deadly nerve gases (poisons that kill by disrupting the victim's nervous system) called sarin and tabun, but they were never used during WWII. After the war the United States, Britain, and the Soviet Union made use of German scientists and captured stockpiles of sarin and tabun to make their own chemical weapons. Chemical weapons were made illegal in 1968, although they have been used since in a few wars. Biological warfare is the use of bacteria and viruses to cause disease in people, animals, or plants. Developments in microbiology have made it possible to "weaponize" many diseases, including anthrax and lassa fever. So far there is no proof that biological weapons have been used in war.

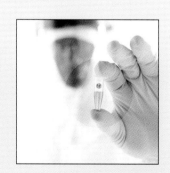

▶ *Tiny amounts of biological weapons could kill large numbers of people.*

▲ *An infantry soldier uses a shoulder-mounted missile to attack an armored target, such as a tank or building.*

adopted and developed in the U.S. Minuteman III missile of 1970. This weapon carried Multiple Independently Targetable Reentry Vehicles (MIRVs), warheads that could maneuver in space, and fly toward a different target. A single ICBM can attack a dozen targets.

Antitank missiles

To pierce tank armor, antitank missiles use shaped-charge warheads, invented in the 1930s. A cone-shaped explosive is detonated inside a thin metal case, creating a jet of extremely hot gas that can punch through thick steel plate. The latest antitank weapons use super-dense uranium shells to punch through thick armor.

SEA WARFARE

▲ *USS* Alabama *was one of the last battleships to see action. Its nine 16-inch (410 mm) guns were fired in combat in the 1980s. Big-gun battleships are no longer used by modern navies.*

Covering two-thirds of the world's surface, the sea has always played an important role in human history, both as a source of food and a means of communication, and the conduct of war at sea is as old as war itself.

The first warships were galleys, a design that was so effective that it was used for 2,500 years. Galleys were powered by a combination of sails and rows of oars, the latter arranged one above the other in banks. In battle the oars made the galley maneuverable and provided the power needed to thrust the ship's main weapon, a ram, into the enemy's hull and sink it.

In ancient times galleys were named for the number of oars they carried, or the number of oarsmen. The first war galleys, built and operated by the Phoenicians (from Lebanon) between 1100 and 800 B.C., were called pentecoters because they had a single row of 50 oarsmen. These small galleys, which were 50 ft (15 m) long, could sustain a speed of 6 knots (7 mph, or 11 km/h) for a short time but were soon replaced by faster Greek galleys, such as the trireme, with its three banks of oars, and the

MILITARY AND SECURITY

larger and even faster septireme, which assigned three or four oarsman to each oar. The design of the galley was developed further still by the Romans, who equipped their ships with catapults and a large plank with a spike on the end called a corvus. This was dropped onto the enemy ship to secure it, while marines boarded and captured the vessel.

▶ The Peloponnesian War in the late 5th century B.C. between Greek city states saw some of the fiercest sea battles in the ancient world.

KEY COMPONENTS

Greek trireme

It was the invention of the trireme, with its three sets of oars, around 500 B.C. that transformed the Greeks into a great seafaring nation. Triremes were usually around 150 ft (46 m) long and carried a crew of 200, of whom 170 were rowers. The latter were not slaves but highly trained oarsmen, and they were grouped in tiers of three. Unlike later vessels, triremes only employed one man per oar. Through this arrangement, the builders of the triremes were able fit a large number of rowers in a relatively small space, making the boats highly maneuverable and able to accelerate quickly.

Triremes could cruise at 6 knots (7 mph, or 11 km/h) for long periods of time and accelerate to 12 knots (14 mph, or 22 km/h) when ramming, making them formidable opponents. The trireme has been described as a ram with a ship attached. It was designed to fit the maximum number of rowers (to propel the craft with great force) in the small

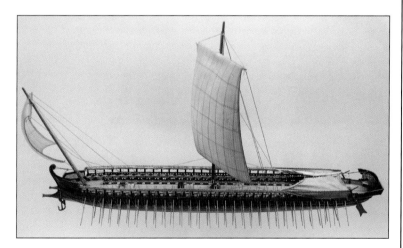

▲ As this model shows, triremes also used sail power for long journeys, but relied on oars during battle. The ship was steered by large oars, or steerboards, attached at the rear of the ship.

amount of space allowed by its ramlike shape. However, the long, thin, top-heavy design made the trireme unstable, so its hull was weighed down with stones. A leather shield protected the rowers from missiles thrown by slings and catapults.

SEA WARFARE

The age of gunpowder

Although galleys continued to be used in war as late as 1790 by the Swedes and Russians, the advent of gunpowder in the 13th century marked the point at which they went into decline. In a world where an opponent could

COPPER-BOTTOMED SHIPS

One of the most significant pieces of metal and ship technology to develop in the 18th century was the technique of covering ships' hulls below the water line with copper sheets. The aim of this was to correct a serious problem faced by all wooden sailing ships—particularly those sailing in the warm waters of the Caribbean and Southeast Asia—the teredo, or shipworm. This creature would literally eat its way through the hull of ships, shortening the life of their timbers. After several ineffective solutions were tried, the idea of using copper to plate boat's hulls was put to the Royal Navy in 1708. The navy rejected it on the grounds of cost, but by 1761 there was an urgent need to find a solution to the problem. Hence, HMS *Alarm* became the first warship to have a copper bottom. The technique was soon in common use on both naval and merchant (commercial) ships.

◀ The copper plate on ships' hulls often went green from corrosion by the salty sea water. Iron and zinc was added to the copper to reduce this problem.

▲ HMS Victory, *the flagship of the British navy in the early 19th century, carried a total of 104 cannons, mostly firing through gunports along the sides.*

keep his distance and bombard the enemy with guns, galleys, with their tactics of ramming and boarding, had no future.

The first true sailing warship was the cog, built in the 12th century. This was the result of two inventions. First came the introduction of the stern rudder, which allowed the ship to be steered effectively in difficult winds. Second, ships were built with deeper and stronger hulls that made them steadier in rougher seas.

In the early 16th century heavily armed galleons came to dominate naval warfare. In 1501 the French invented the gun port. This was a hatch or windowlike opening in the side of the ship that could be opened in order to fire

MILITARY AND SECURITY

the guns and closed in bad weather. This allowed guns to be placed below decks. The result was that a far greater number of guns could be placed aboard sailing ships. English sailor Sir John Hawkins (1532–1595) removed the high front and back of the galleon to create a steadier vessel, or ship-of-the-line, which could carry several decks of cannons. These ships did maximum damage when all the guns on one side were fired together (called a broadside). This design remained the basis for warship construction until the 1870s.

Steam and steel

Industrial development in the 19th century inevitably revolutionized warship design. In 1783 the French engineer, the Marquis de

---

CONGREVE ROCKETS

The idea of cities being bombarded by rockets may seem a modern one. However, in 1805 the British navy began using Congreve rockets to attack ports. The rockets were developed by Sir William Congreve (1772–1828), an army officer who had seen the weapon in action when fighting rocket-armed forces in India in the 1790s. Encased in metal, Congreve's rockets were up to 3 ft (1 m) long, and weighed around 32 lb (15 kg). In the War of 1812, the British navy attacked several U.S. ports with these rockets, remembered in the "the rockets' red glare" lyric in the U.S. national anthem.

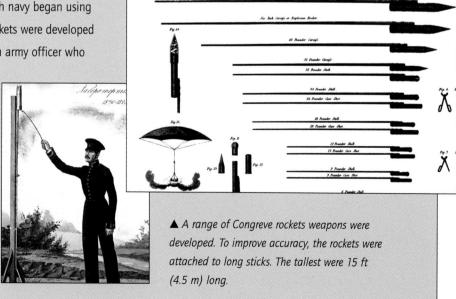

▲ A range of Congreve rockets weapons were developed. To improve accuracy, the rockets were attached to long sticks. The tallest were 15 ft (4.5 m) long.

43

SEA WARFARE

Jouffroy (1751–1832) built the steam-propelled *Pyroscape*, which was driven by a paddle at the back of the ship. Although others soon copied his idea, these ships were unsuited to the rough open seas. The first steam warship, the *Demologus*, was designed in 1813 by the U.S. engineer Robert Fulton (1765–1815). However, the paddle used to propel the ship was vulnerable to enemy gunfire. Later warships were powered by screw propellers, although steamships continued to use sails as an additional source of power until the 1870s.

Another impact of the Industrial Revolution was the invention of warships with iron and, later, steel hulls. The idea of using iron plates to protect the decks of wooden hulls of ships had been used as early as the 1590s by the Korean

▲ *The iron-hulled steamship HMS* Warrior, *seen here in 1864, dwarfed the Royal Navy's wooden warships.*

navy. In 1860 the British took the idea further when HMS *Warrior* became the the first warship that had a hull made solely of iron. The *Warrior* was 480 ft (145 m) long, carried 48 smoothbore guns on her broadside and had a speed of 14 knots (17 mph, or 26 km/h).

In 1862 US shipwright John Ericsson made a key contribution to the development of warships. With the launch of *Monitor* he radically changed the way guns were placed on warships. The vessel's guns were positioned in a rotating turret on the deck of the ship, which allowed them to fire at targets to the front and rear of the ship, not just the side.

MILITARY AND SECURITY

THE SCREW PROPELLER

Although John Ericsson had invented the screw propeller in 1836 and fitted it the to U.S. warship *Princeton* in 1843, navies around the world were reluctant to adopt it. It was argued that although it was not as vulnerable as the paddle, it lacked power. The argument was finally resolved in a dramatic manner in 1845. The British navy lashed the paddle wheeler *Alecto* and the screw propeller *Rattler* together in order to see which was the stronger. On April 3 in the North Sea the contest began. Initially, there was little movement as each vessel increased the power. Then, slowly the *Rattler* inched forward, gathering speed until it was towing the 900-ton (815 metric ton) *Alecto* backward. There was no more argument over the power and superiority of the screw.

▲ A brass screw propeller is one of the displays at the naval museum at Camden, New Jersey.

KEY COMPONENTS

Gloire

In 1859 the French built *Gloire*, which was the first warship with a wooden hull covered in iron. Such ships were called ironclads. Designed by Dupuy de Lôme, the iron cladding was added as protective armor for the French frigate. This covering stretched from a point 6 feet (nearly 2 m) below the waterline to the upper deck and was 4.75 inches (12 cm) thick. The layer was completed by a 26-inch (66 cm) oak backing.

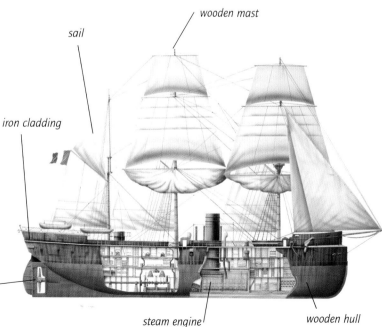

45

SEA WARFARE

In 1871 Edward Reed combined a steam engine, a steel hull, and the massive firepower of 12.5-inch (30 cm) guns mounted in turrets by launching the 14,000-ton (12,700 metric ton) HMS *Devastation*—the first truly modern battleship. Battleships continued to grow in the early 20th century, the largest being built by the Japanese in the 1930s. However, these huge warships were vulnerable to air attack and by the 1940s were eclipsed by the aircraft carrier.

The era of the aircraft carrier

The first time that an airplane was flown from a ship was November 14, 1910, when Eugene B. Ely took off from the U.S. cruiser *Birmingham*. The following year Ely became the first person to land an aircraft on a ship. The first ships used as aircraft carriers were simply modified warships with a flat deck placed on their backs. HMS *Argus*, launched in 1918, had a full unobstructed flight deck and was the first modern aircraft carrier. In subsequent years the British Royal Navy led the world in the technical design of carriers.

To assist landing, rows of wires were placed across the landing deck to snag hooks trailing from the aircraft and stop them rolling off into the sea. In 1920 the *Eagle* was the first ship to

SOCIETY AND INVENTIONS

Clash of the ironclads

In 1862, two ironclad warships went head to head for the first time at the Battle of Hampton Roads, taking place along a wide waterway in Virginia. The battle was the most important naval engagement of the US Civil War. The Confederate ship *Virginia* was armed with 12 guns, which were fired through portholes positioned along the ship. The Union vessel *Monitor* had just two guns, but they were encased in a turret that swiveled side to side. The battle lasted two days and ended as a draw. Neither ironclad's guns could penetrate the other's armor.

▼ *Despite being out-gunned, the* Monitor *(center) matched the* Virginia *(left) for firepower thanks to its rotating gun turret. Within decades, all battleships used similar turrets.*

MILITARY AND SECURITY

KEY COMPONENTS

Dreadnoughts

The introduction of HMS *Dreadnought* in 1906 proved to be a watershed in the history of battleship design. The impact of the vessel was so great that battleship design is divided into two eras: pre- and post-dreadnought. Not only was it the first warship to have steam turbine propulsion, it was also the first "all-big-gun" ship in the world. Previously, battleships had tended to be armed with both a main battery of high-caliber guns and supporting batteries of smaller weapons. *Dreadnought's* main armament was all of one caliber, consisting of ten 12-inch (305 mm) guns in five turrets. The 1914 battleship shown here, the USS *Texas* (now a museum ship near Houston) is a super-dreadnought. It carried ten 14-inch (360 mm) caliber guns housed in five twin turrets. *Texas* also had 21 smaller 5-inch (130 mm) guns.

have its bridge and funnels set to one side of the deck to create more room for landing. In 1951 the use of heavier jet fighters led the British to invent a steam catapult, which is used to assist the jet as it takes off by giving an extra boost of power. In 1961 the U.S. Army launched the first supercarrier—the 84,000-ton (76,000 metric ton) *Enterprise*.

Underwater weaponry

Technological developments have allowed war to be fought under the sea, as well as on and above it. U.S. inventor David Bushnell (1742–1824) devised the mine, an underwater explosive, in 1775. The torpedo, meanwhile, was the joint invention of the Englishman Robert

FACTS AND FIGURES

- At over 80,000 tons (72,000 metric tons), the Japanese *Yamato* was the heaviest battleship ever seen. The WWII warship was 862 ft (263 m) long and had a crew of 2,500. Among its formidable array of weapons were nine 18-inch (46 cm), twelve 6-inch (15 cm), and twelve 5-inch (13 cm) guns. Its armor was designed to withstand a 2,200-lb (1,000 kg) bomb dropped from a height of 10,000 ft (3,000 m). Despite this, the *Yamato* was sunk by U.S. airplanes on April 7, 1945.

- The record speed any kind of military vessel is 91.9 knots (110 mph, or 170 km/h) set by the U.S. attack hovercraft SES100B on January 25, 1980.

SEA WARFARE

Whitehead (1823-1905) and the Austrian Giovanni Luppis. In 1866, Whitehead built an underwater missile that was 14 ft (4 m) long and driven by compressed air at a speed of 6 knots (7.2 mph, or 11 km/h) for 700 yards (640 m). Although widely used in World War I (1914-1918), technical problems made torpedoes unreliable. The problems were overcome in the 1930s by the Japanese, who invented the Long Lance torpedo with a speed of 49 knots (56 mph, or 91 km/h) and a range of 11 miles (18 km). Later weapons had electronic guidance systems.

The most dangerous underwater weapon is the submarine. In 1578 the Englishman William Bourne first put forward the idea, but it was the Dutchman William van Drebble who built the first one on the Thames River in 1620. In 1776 David Bushnell's *American Turtle* became the first military submarine.

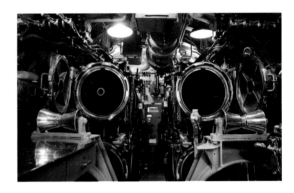

▲ *The torpedoes are fired from water tight tubes. Once loaded the inner doors are locked and outer ones opened to flood the tubes, making the torpedo ready to fire.*

AMERICAN TURTLE

The world's first ever military submarine was created by David Bushnell, who began work on the design of the vessel while still a freshman at Yale in 1771. The American Revolution gave Bushnell the opportunity to put his ideas into practice, and in 1775 he created the *American Turtle*. The *Turtle* was a one-man submarine driven by a hand-operated propeller. The operator of the boat was also responsible for steering and letting water in and out of the ship's tanks for ballast (weight). The idea was that the *Turtle* would creep up under British warships, attach kegs of gunpowder to their hulls, and then retreat to a safe distance to watch the resulting explosion. Unfortunately for Bushnell, the *Turtle* proved unsuccessful in battle: its attack on the British man-of-war HMS *Eagle* failed when the gunpowder proved incapable of damaging the ship's hull.

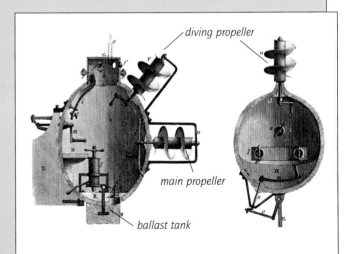

▲ *The* American Turtle *was a large wooden barrel with a brass viewing compartment at the top. The vessel submerged when water was pumped into the ballast tank. The operator then went deeper by cranking the diving propeller.*

MILITARY AND SECURITY

▶ This WWII submarine has two torpedo tube doors on the bow. The submarine was steered underwater using adjustable diving planes. On the surface, the captain could control the vessel from the conning tower.

KEY COMPONENTS

Supercarrier

Aircraft carriers are the largest warships afloat today. The U.S. Navy's Nimitz class are supercarriers. They can carry 80 aircraft and have a crew of more than 5,000.

▶ Supercarriers spend six months at sea, generally escorted by a battle group of smaller ships, which are there to protect the carrier from attack.

49

SEA WARFARE

It was the invention of electric and fuel motors in the 1880s and the torpedo that gave submarines both weaponry and an effective means of propulsion. However, submarines had to rise to the surface to recharge the electric batteries used to power the craft underwater. Diesel engines were used to power the submarine at the surface at speeds of up to 10 knots (12 mph, or 19 km/h) while recharging their batteries ready to submerge again.

▶ *Fighting ships seldom see enemy vessels in modern naval engagements since they are attacked by ship-borne aircraft, hidden submarines or with missiles fired from far beyond the horizon.*

In 1955, *Nautilus*, the first nuclear-powered submarine, was built. *Nautilus* could sail right around the world while submerged. The addition of nuclear ballistic missiles in the 1960s, meanwhile, created submarines capable of destroying cities thousands of miles away.

SCIENTIFIC PRINCIPLES

Nuclear-powered submarines

Nuclear submarines are powered by a pressurized-water reactor (PWR). In a PWR a uranium-based nuclear reactor heats a primary circuit of water, which is pressurized to prevent it from boiling. This in turn heats a steam generator that marks the start of the secondary circuit. The water in this is transformed into steam, which drives a turbine that is directly connected to the submarine's propeller. The steam then travels through a series of condensers, which turn it back into water. This then flows back into the steam generator, and the whole process starts again.

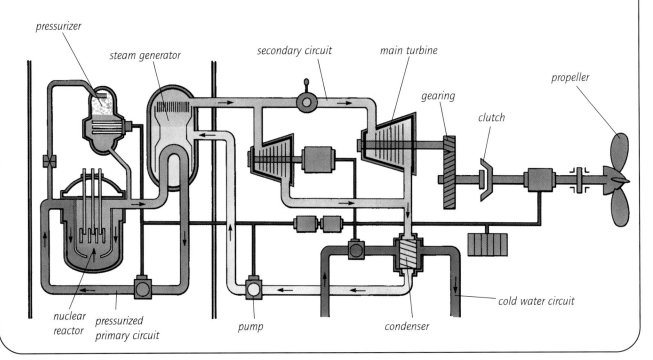

MILITARY AND SECURITY

SOCIETY AND INVENTIONS

Submarines and the nuclear age

The advent of the nuclear-powered, nuclear-armed submarine in the second half of the 20th century had a vast effect that stretched far beyond the realms of naval warfare. The new nuclear-powered vessels are capable of traveling virtually indefinitely without surfacing—until food supplies run out—and are thus very difficult to detect. Since they are often armed with nuclear missiles with 2,500-mile (4,000 km) ranges, no target in the world is beyond their range. Submarine-based nuclear arsenals are almost completely beyond attack by enemies and that ensures that if one country launches a nuclear attack, a just as devastating strike will be made in retaliation.

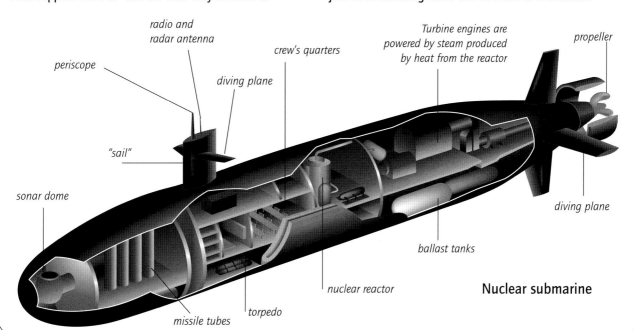

Nuclear submarine

LOCKS AND KEYS

People have always needed to protect their possessions and have developed security technology that ranges from locks and keys to impenetrable vaults.

One of the earliest known locks was found in the ruins of the palace of Khorsabad in Iraq. Made entirely of wood, this lock is probably 4,000 years old. The door is secured with a sliding wooden bolt. The lock mechanism, called a pin-tumbler, consists of several holes drilled through

▲ *Every lock in a building needs a different key, carried together on a key ring. Janitors often have a master key that opens every lock in their buildings.*

the bolt with matching pins attached to the door above. As the bolt is slid across the door, the pins drop into the holes, locking it in place. The key is a large wooden bar with a group of upright pegs that matches the pattern of holes. When the key is inserted, the pegs lift the pins out of the bolt, and the door is unlocked. Locks of this type are still used in parts of North Africa and the Middle East, and the falling pin design forms the basic principle of many locks, including the famous Yale lock.

Ancient Greek and Roman locks

The ancient Greeks were the first to use metal keys, but their locks were less sophisticated than the older pin-tumbler designs. The bolt was

FACTS AND FIGURES

- Mortise locks have the locking mechanism mortised (cut) into the door. They are very strong locks when installed in sturdy doors.
- Rim locks, including their bolts, are enclosed in a metal case that is mounted on the inside of the door.
- Cylindrical (or knob) locks form part of the doorknob—the keyhole is actually set in the doorknob (below).

MILITARY AND SECURITY

▶ A lock is only as strong as the door it closes. Castles needed stronger defenses than that, and many gateways were secured by an iron grill, called a portcullis, that was raised to let people in, and lowered onto uninvited visitors.

moved by an L-shaped key made from iron. The key was passed through a hole in the door and turned, engaging the bolt and drawing it back. This type of lock provided little security because it was not encoded in any way. Anyone who could fashion an L-shape of the right dimensions could open the door.

Roman locks were the first to be made entirely out of metal, usually iron for the lock and bronze for the key. The Romans invented wards, which are intricate patterns of matching

SOCIETY AND INVENTIONS

Harry Houdini

Ever since the invention of locks people have devised ways of defeating them. Perhaps the greatest lock-picker of all time was U.S.-Hungarian magician Harry Houdini (1874–1926). Houdini specialized in entertaining thousands of people with his sensational escape acts. In a typical act he would be shackled with metal chains and locked into a wooden box. The weighted box was then submerged, leaving him only a few minutes to break free before he was drowned by the rising water. In addition to his extraordinary skill at manipulating locks, Houdini relied on great physical strength and agility to make his escapes. He exhibited his skills in a series of movies made between 1916 and 1923.

▶ Harry Houdini prepares to escape from locked chains in 1899.

53

LOCKS AND KEYS

slots on both key and lock. Once the key is slotted onto the ward, it can be rotated around the ward rail to operate the bolt. Warded locks have always been relatively easy to pick because special keys could be made to pass around the ward even if they did not fit perfectly.

Medieval locks

During the Middle Ages a great deal of skill was used in crafting locks, but the technology changed little. Keyholes were often hidden behind secret shutters to confuse the picker, and false keyholes were often cut into the door to waste would-be thieves' time. However, security still depended on complex warding.

One innovation was the English letter-lock of the 17th century. A number of rings inscribed with letters or numbers were threaded onto a spindle. When the rings were turned to form a particular number or word, slots inside the rings were aligned and the spindle could then be drawn out.

Early modern locks

It was not until the 19th century that any real advances in lock design were made. In 1778 English locksmith Robert Barron patented a double-acting tumbler. A tumbler is a lever that falls into a slot and cannot be moved unless it is raised to a set height. Inserting the key raises the tumblers, while turning it slides the bolt out. The Barron lock had two tumblers, and the key had to

▼ *A simple lock comprising a bolt and latch. The latch is a hook that can be worked from both sides of the door, while a bolt locks the door from the inside only.*

> ## BRAMAH'S LOCK
>
> Joseph Bramah's 1784 patent lock used a key that had lengthwise slots of varying lengths cut into the end. When the key was pushed into the lock, it depressed a number of internal mechanisms called slides to specific depths depending on the length of the slots on the key. Only when all the slides were depressed to the right depth could the key be turned and the bolt drawn. Bramah's lock and key worked on the same principle as the D-shaped locks now used to secure bicycles.

MILITARY AND SECURITY

raise each by a different amount before the bolt would slide. The principle behind this design is still used in all lever locks today, but a determined lock picker could still open Barron's lock.

In 1784 an entirely new lock was patented by English engineer Joseph Bramah (1748-1814). It used a small key yet provided a level of security never achieved before. To manufacture his locks, Bramah constructed a geared lathe, one of the first metalworking machines designed for mass production. The locks were very expensive, but Bramah was so sure of their security that he offered a reward of £210—a huge sum in those days—to the first person who could pick one. This challenge stood for 67 years until a U.S. locksmith, A.C. Hobbs, succeeded and claimed the reward. However, picking the lock took Hobbs more than 50 hours!

PADLOCKS

Padlocks are removable locks with a hinged or pivoting curved bar called a shackle, which can be used to join two objects together. The locking mechanism is enclosed in a steel case. To shut the lock, the shackle is simply pushed into a hole in the steel case. Inside, a lever falls into a slot at the tip of the shackle, holding it in place. A combination or a key is used to release the shackle and open the lock.

COMBINATION LOCKS

The combination lock, based on 17th-century English letter-locks, consists of a number of rings attached to a central spindle. When the correct combination is entered, the spindle can be drawn out because slots inside the rings are aligned. A combination lock with four rings and 100 numbers on each ring offers a picker 100 million different options. Because there are many possible combinations and no keyholes in which an explosive charge can be placed, the combination lock was the most secure form of protection for safes and strong rooms in the 19th century. An added bonus is that combinations can be easily changed. Some lock pickers have used stethoscopes to listen for telltale clicks as the rings inside the combination lock fall into line.

55

LOCKS AND KEYS

▶ Modern car keys do not need to be put in the lock at all. They send a radio signal to the car, which activates the doors' magnetic locks.

Yale locks

By the middle of the 19th century the lock industry was booming. The fast-growing economies of the United States and Europe provided huge demand for locks, and hundreds of new types of lock were patented. Most of these, however, were merely variations on existing lock designs.

In 1848, Linus Yale (1821–1868) patented a pin-tumbler lock based on ancient principles. In the 1860s he designed the Yale cylinder lock, with its familiar serrated key. Pins are raised to the correct heights by the serrated edge, thus releasing the cylinder, which can then be turned. This lock is now widely used in the doors.

Time locks

In the 1870s a new crime wave swept across the United States; armed robbers would force bank cashiers to open their safes. To combat this, U.S. inventor James Sargent built the first time lock in 1873. His lock contained a built-in clock and could only be opened at a preset time.

THE GOLD VAULT

The vault beneath the Federal Reserve Bank of New York contains gold reserves worth $86 billion and must be one of the most secure places on Earth. Made of steel-reinforced concrete, the three-story bunker is sunk 80 ft (27 m) below street level, where it rests on the bedrock of Manhattan Island. There are no doors into the gold vault. Entry is through a 10-foot (3 m) passageway cut through a 90-ton (80 metric ton) steel cylinder that revolves vertically in a 140-ton (127 metric ton) steel-and-concrete frame. To seal the vault, the cylinder, which is slightly tapered, is rotated through 90° and lowered into the frame, like pushing a cork into a bottle. Eight large bolts are then inserted into the cylinder to secure it. Bank personnel can open the vault by unlocking a series of time and combination locks, but no one person has all the combinations needed. In addition to these measures, the vault is protected by guards and an electronic surveillance system. No one has ever tried to steal the gold.

▲ The walls around a bank vault or strongroom are too thick to cut through or blow a hole in.

MILITARY AND SECURITY

KEY COMPONENTS

Types of locks

1 A warded lock relies on a specific matching pattern of slots and grooves on both lock and key. Once inserted, the correct key can be slid onto the ward rail and rotated, drawing back the bolt.

2 A Yale cylinder lock uses the pin-tumbler principle. The pins are cut into two parts of varying lengths. Notches cut into the flat key raise each pin to a specific height at which the cut between the two parts of the pin aligns with the edge of the cylinder. The cylinder can then be rotated, releasing the bolt.

3 A combination lock consists of a series of rings threaded onto a central spindle. A slot is cut into the outer edge of each ring, and a raised stud is present on the inner surface. When the outer knob is rotated, the pin engages an arm on the spindle, which in turn engages the stud on the first ring. The rings are connected to one another by the same mechanism. By turning the knob through a specific series of clockwise and counterclockwise rotations, using the numbers etched onto the dial surface as a reference, the slots on the rings are aligned, allowing the bolt to be drawn out.

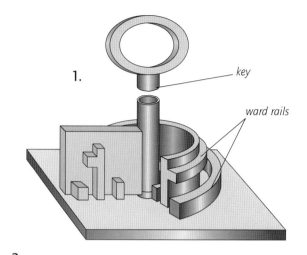

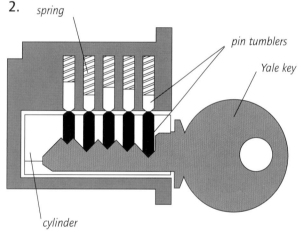

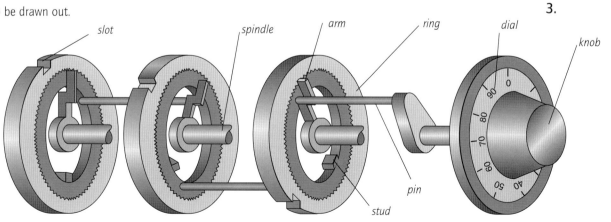

LOCKS AND KEYS

▲ Hotels often use card keys, which contain a small microchip (seen here as a gold square). The chip is programmed to open just the lock of one room. If the card key is lost by a guest, another one can be programmed.

Modern locks and keys

The importance of locks as a protection against bank robbers declined after World War II (1939–1945), during which the knowledge of explosives became widespread. Locks became more difficult to pick, and thieves usually ignored them and simply blasted them off.

Locks have been invented with many bizarre designs. Some resist explosives, becoming useless if tampered with; others shoot, stab, or seize the hands of intruders. Locks have been made that can be locked with one key but only opened with another. The basic types remain the warded, the lever, and the pin-tumbler, although many variations have been made, often combining the best features of each.

In buildings with many different locks, such as offices, schools, and prisons, it can be very useful to have a "skeleton key" that can open all the doors in the building. This master key can be shaped to avoid the warding of all the locks, or there may be two keyholes in the lock, one for the normal key, one for the master key. Alternatively, the locks can be fitted with two

RESTRAINING DEVICES

Handcuff locks have a row of teeth cut into one side of an open metal loop that can be pushed past a spring-loaded catch until the cuff fits tightly around the wrist. The catch prevented the cuff from being loosened unless the correct key was inserted. Disposable handcuffs, which do not include locks, are now available. Made of strong nylon, they can only be removed by being cut off. These lightweight restraints are particularly useful to police in crowd situations such as riots.

MILITARY AND SECURITY

▶ Computer security keeps unapproved people locked out of the system by asking for a password. The password system is not new. It was used by Roman guards, who challenged every visitor for that day's "watchword."

sets of tumblers and levers or, in the case of Yale locks, two concentric cylinders—one operating inside the other.

Today, like most other forms of technology, locks have become computerized. The code is no longer physically encrypted on a key but is magnetically or digitally encoded on a card. Some locks can even recognize an individual's retina (the inner surface of the eye) or fingerprints. One of the ultimate locks, used on the U.S. nuclear missile launch system, can only be activated by two operators putting different keys into keyholes at opposite ends of the wide control console and turning them at precisely the same time. If one trusted operator was to overpower the other, he would still be unable to launch missiles. He would not be able to reach far enough to turn both keys simultaneously.

SCIENTIFIC PRINCIPLES

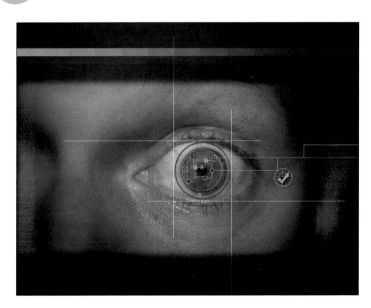

▲ A biometric retina scan takes a picture of the retina and looks for a match among master copies stored in its memory.

The unpickable lock?

It is possible to identify an individual without any risk of error by measuring particular biological features of the human body—a technique called biometry. The most reliable biometric lock developed so far works by scanning the retina (inner surface of the eye) with a laser. Each person has a unique pattern of blood vessels on their retina, in the same way that every person has fingerprints. The biometric lock contains a computer that can recognize the retinas of people who are allowed to enter. There are no keys that could fall into the wrong hands, nor is there any combination to remember, making this one of the most secure forms of lock.

TIMELINE

400,000 B.C. The first evidence of wooden spears found in Europe, used by Neanderthals.

40,000 B.C. The atlatl is invented in North Africa. The weapon throws a dart in a similar way to a bow.

23,000 B.C. Throwing sticks are used in Europe and Africa, similar to the boomerangs of Australian Aboriginals, although most do not come back to the thrower.

20,000 B.C. The first arrowheads date from this time, although no bows of this age are known.

c. 8000 B.C. In Europe flint mines are being worked to make cutting blades.

3500 B.C. Bronze is in use. Wheels are added to sleds to make the first carts.

3,000 B.C. The composite bow, a small but powerful weapon, is invented in northern Asia.

2000 B.C. The first lock is invented in Iraq.

1500 B.C. Iron smelting is first carried out on a large scale. The Egyptians make body armor by attaching small pieces of bronze to leather garments.

500 B.C. Trebuchet catapults are developed in China. The ballista crossbow device is invented in Greece around the same time.

A.D. 1000 First known formula for gunpowder appears in a Chinese manuscript.

1100s The first true sailing warship, the cog, is built in northern Europe.

1400s Metal guns are used for the first time, first in China and India and then Europe.

1415 Longbow revolutionizes warfare in Europe.

1501 Gunport invented for firing cannons from lower decks of ships.

1771 First military submarine, the *American Turtle*, puts to sea.

1805 Congreve rockets used in naval bombardments of harbors.

1813 *Demologus*, the first steam-powered warship is launched.

1835 Samuel Colt patents the Colt revolver.

1836 Francis Smith and John Ericsson come up with a successful design for a screw propeller.

1848 The breech-loading rifle, invented by Johann Dreyse, comes into service.

1860 HMS *Warrior*, the first iron warship, is built in Britain.

MILITARY AND SECURITY

1945 The first atomic bombs are dropped by the United States on the Japanese cities of Hiroshima and Nagasaki.

1952 The first hydrogen bomb is tested by the United States.

1953 The United States develops a surface-to-air missile (SAM) by this time.

1955 USS *Nautilus*, the first nuclear-powered submarine, is launched and sails around the world underwater.

1959 The hovercraft is invented by Christopher Cockerell.

1983 Lockheed builts the F-117A, a stealth fighter that is almost invisible to radar defenses.

2002 Lasers are used to destroy artillery shells.

2007 Estonia suffers a cyberattack from Russia, turning off many of the country's Internet and telecommunication services.

2008 The first plasma weapon is tested, firing balls of super-hot gas at targets. Fighter jets are also fitted with the first laser weapons around the same time.

1860s The first Yale locks are manufactured.

1862 First engagement between ironclad ships takes place at Battle of Hampton Roads.

1875 Chemist Alfred Nobel invents dynamite.

1884 Hiram Maxim develops the first fully automatic machine gun.

1902 Trinitrotoluene (TNT) is invented in Germany.

1918 First aircraft carrier, HMS *Argus*, is launched.

1939 The first jet-powered airplane, the Heinkel He-178, is flown. The first practical helicopter is flown in the United States by engineer Igor Sikorsky.

1942 Enrico Fermi sets up the first self-sustaining nuclear chain reaction.

1943 The Colossus computers are developed in England to crack German military codes.

1944 The V-1 flying bomb, or "doodlebug," becomes the first cruise missile.

GLOSSARY

American Revolution (1775–1783) A war fought between Great Britain and the 13 British colonies in North America. The colonies were victorious and went on to form the United States.

ancient Greece A civilization that existed on the mainland and islands of modern-day Greece and Turkey between 2000 and 300 B.C.

artillery Large guns designed to fire heavy projectiles such as explosive shells. Artillery pieces can be mobile, stationary, or mounted on ships or airplanes and are often capable of destroying targets many miles away.

atom The smallest unit of a chemical element that can take part in a chemical reaction. Atoms are composed of a central nucleus made up of protons and neutrons surrounded by shells of electrons.

bacteria Single-celled microorganisms that are present almost everywhere on Earth. Bacteria can only be seen through a microscope, and most are spherical, rodlike, or spiral in shape. Some bacteria are beneficial to people, helping us digest our food and playing an important part in the preparation of some foods. Others, however, are responsible for serious diseases such as cholera.

Civil War (1861–1865) Also called the American Civil War, this conflict was fought between the Northern (Union) and the Southern (Confederate) states. One of the major disputes was over slavery, which the South supported and the North wanted to stop. The North won, and slavery was made illegal throughout the United States.

Cold War An extended period of hostility that ran from 1945 to the start of the 1990s. A group of communist states, led by the Soviet Union (a now dissolved communist empire centered on Russia), formed one side, and they were opposed by a group of noncommunist nations, dominated by the United States. Although conflict never broke out, there was a constant threat of war, and both sides produced vast numbers of nuclear and conventional weapons in case of attack. The Cold War ended with the collapse of many communist governments in the late 1980s and early 1990s.

digital A measuring system, such as binary code, that can store information as a number of discrete values. A binary computer is an example of a digital device since it uses only two values—one and zero—to store data.

gamma rays High-energy electromagnetic radiation with a wavelength shorter than 0.01 nm (billionths of a meter). Gamma rays are emitted after nuclear reactions.

Holocaust The systematic murder of six million Jews and millions of gypsies, Poles, and Russians by the Nazis during World War II. Victims of the Holocaust were executed, starved, or worked to death, often at specially constructed concentration camps.

Industrial Revolution A great change in social and economic organization brought about by the replacement of hand tools by machines and power tools, and the development of large-scale industrial production methods. The Industrial Revolution started in England around 1760 and spread to the rest of Europe and the United States.

infrared A type of electromagnetic radiation covering wavelengths between 0.75 and 10,000 μm (millionths of a meter). Infrared rays have a strong heating effect and are emitted by hot objects.

Korean War (1950–1953) A war in which a U.S.-dominated United Nations coalition came to the aid of South Korea during an invasion by North Korea, which was aided by the Soviet Union and communist China. The war was ultimately indecisive, but was one of the key military confrontations of the Cold War.

medieval Of or from the Middle Ages.

metallurgist A person who studies the science of metals, particularly the extraction and processing of metal ores.

Middle Ages A period of European history that ran from around A.D. 500 to c. 1450.

Nazi The name of a political party or its members that, led by Adolf Hitler, ruled Germany from 1933 to 1945. The Nazis suppressed all opposition and built up Germany's military strength. They were the main protagonists of World War II. See also Holocaust.

neutron A component of an atom's nucleus. Neutrons do not have an electric charge.

orbit The path, shaped like a circle or an ellipse, that an object in space takes around another object.

radiation The act of giving off radioactive particles, heat, or electromagnetic waves.

radio waves A type of electromagnetic wave with a wavelength between 1 mm and 1,000 m. Radio waves are used by people to transmit information over long distances.

FURTHER RESOURCES

radioactivity The disintegration of atomic nuclei accompanied by the giving off of particles or electromagnetic waves.

Rome The ancient civilization that began in the Italian city of Rome around 700 B.C. and had established a vast empire around the Mediterranean by 200 A.D. The Romans are noted for being the first to bring law and order to Europe and for their great works of engineering. Roman—noun, adjective

satellite A natural or artificial object in orbit around a star or planet.

Soviet Of or from the USSR, a communist state that existed from 1923 to 1989 and included present-day Russia.

telecommunications Sending messages over a distance, usually involving electrical signals or electromagnetic waves.

Vietnam War (1957–1975) A conflict between communist North Vietnam (supported by the Soviet Union and China) and noncommunist South Vietnam (supported by the United States). U.S. soldiers were actively involved in the war from around 1965 to 1973, after which hostile public opinion in the United States forced their withdrawal. North Vietnam conquered the South in 1975.

virus A tiny, disease-causing particle that consists of protein combined with genetic material (DNA or RNA—ribonucleic acid). Viruses are only capable of replicating inside living cells, and for this reason many scientists do not consider them to be living organisms. Other scientists, however, consider them to be a type of microorganism.

World War I (1914–1918) A war fought mainly in Europe between the Central Powers—Germany, the Austro-Hungarian Empire (present-day Austria and Hungary), and the Ottoman Empire (now Turkey)—and the Allies: France, the British Empire, Russia, and the United States. The Allies eventually won the conflict, but millions of soldiers on both sides lost their lives.

World War II (1939–1945) The most destructive conflict in history, fought mainly in Europe, East Asia, and North Africa. The Axis powers (Germany, Austria, Japan, and Italy) were opposed by the Allies (Britain, the United States, France, and the Soviet Union). Germany surrendered in April 1945, but Japan fought on until August, when atomic weapons dropped by U.S. aircraft destroyed the Japanese cities of Hiroshima and Nagasaki.

FURTHER RESOURCES

Books

The Archaeology Of Warfare: Prehistories of Raiding and Conquest. Gainesville: University Press Of Florida 2006.

Arms & Armor by Michele Byam. New York: DK Pub., 2004.

Conway's Battleships: The Definitive Visual Reference To The World's All-Big-Gun Ships. Annapolis, MD: Naval Institute Press, 2008.

The Story Of Guns: How They Changed The World by Katherine Mclean Brevard. Minneapolis, MN: Compass Point Books, 2010.

Websites

History Channel: Inside WWII
http://www.history.com/interactives/inside-wwii-interactive

Discovery Channel: Great American Planes
http://www.yourdiscovery.com/flight/great_american_planes/index.shtml

Howstuffworks: Technology of War
http://science.howstuffworks.com/war-tech.htm

British Museum: Sparta
http://www.ancientgreece.co.uk/sparta/home_set.html

INDEX

A
aircraft carrier 46, 49
AK-47 20
American Civil War 18
American Turtle submarine 48
ammunition 17-18, 22
armor 4-7, 12, 15, 29, 39, 45-47
artillery 10, 20-24, 26, 28, 32
ax 4-6, 12, 16

B
ballastite 22-24
battering ram 8, 9
Battle of Hampton Roads 46
battleships 40, 46-47
bayonet 16
biological warfare 39
biometric security 59
Blitzkrieg 28
Bronze Age 4
bullet 17, 21, 25

C
cannons 14, 15, 42, 43
castle 8
catapult 10-11, 47
catapults 10, 41
cavalry 10-12, 14-15
chariot 10
chemical weapons 27, 29, 39
Chinese weapons 7, 12-13
chivalry 13
composite bow 6-7
crossbow 6-7, 11-12
cruise missiles 35

D–G
dreadnought 47
dynamite 22-23
early warning radars 34
flamethrowers 24
flintlock musket 16, 17
gas masks 28-29
Gatling, Richard 24
Global Positioning System (GPS) 34
grenade 13, 25-27
guided missiles 32, 34-36, 38
gun 14, 16-18, 20-22, 24-26, 28-29, 40, 42, 46, 47
guncotton 22-23
gunpowder 12-15, 19, 22-23, 42, 48

H
Haber, Fritz 27
handcuff 58
handguns 14-15, 17
helmets 4
HMS *Warrior* 44
Holocaust 29
hydrogen bomb 30, 31

I
Intercontinental ballistic missiles (ICBMs) 33, 36-37, 39
Industrial Revolution 18, 44
infantry 10-12, 14-16, 18, 20-21, 25-26, 28, 39
ironclads 45-46

K
Kalashnikov, Mikhail 20
key 9, 11, 17, 25, 35, 37, 41, 44-45, 47, 49, 52-59
knights 12-13

L
lock 52-59
 card 58-59
 combination 40, 55-57, 59
 cylindrical 52
 double-acting tumbler 54
 mortise 52
 rim 52
 warded 54, 57-58
 Yale cylinder 57
longbow 6-7, 12

M
maces 4
machine gun 24-26
Manhattan Project 31
matchlock musket 15-16
Maxim, Hiram 24, 25
military satellites 34
mines 12, 27
missile 6-7, 33-39, 48, 51, 59
 guidance systems 38
mortar 18, 25-27

N
Nagasaki 30
napalm 24
nitroglycerin 22-23

Nobel, Alfred 22-23
nuclear weapons 30-33
nuclear-powered vessels 50-51

P
padlocks 55
percussion cap 17
pike 12, 16
plastic explosives 23
poison gas 27

R
rifle 16, 18-2, 23
 automatic 19-20, 24
 bolt-action 19-20
 breech-loading 18-21
 muzzle-loading 18-19
 self-loading 19-20
rockets 12, 28, 34-35, 37, 43
Roman army 5

S
saddle 11-12
samurai 4
shrapnel 15
sieges 8-9, 14, 18
stirrup 10-12
Strategic Defense Initiative 33
submarine 36, 48-51
surface-to-air missiles (SAM) 35-36
sword 4-6, 12, 16

T
tank 24, 27-29, 39, 48
TNT 23, 31
torpedo 47-51
Trident II missile 37
trireme 40-41

V
V-1 flying bomb 35
vault 56
Vietnam War 24

W
war cemetery 26
Winchester rifle 18
World War I 20, 21, 24-25, 27-28, 32, 39, 48
World War II 20, 23-24, 26-30, 32, 35, 39, 47, 49, 58